Partial Differential Equations in China

Mathematics and Its Applications

Managing Editor:

M. HAZEWINKEL

Centre for Mathematics and Computer Science, Amsterdam, The Netherlands

Volume 288

Partial Differential Equations in China

edited by

Chaohao Gu
Fudan University,
Shanghai, China

Xiaxi Ding
Institute of Applied Mathematics,
Academia Sinica,
Beijing, China

and

Chung-Chun Yang
Department of Mathematics,
The Hong Kong University of Science and Technology,
Clear Water Bay,
Kowloon, Hong Kong

SPRINGER SCIENCE+BUSINESS MEDIA, B.V.

A C.I.P. Catalogue record for this book is available from the Library of Congress.

ISBN 978-0-7923-2857-5   ISBN 978-94-011-1198-0 (eBook)
DOI 10.1007/978-94-011-1198-0

Printed on acid-free paper

TABLE OF CONTENTS

Preface   vii

Contributors   ix

Part One: Survey Papers

Degenerate Parabolic and Elliptic Equations   1
     Yazhe Chen

Nonlinear Hyperbolic Conservation Laws   19
     Xiaxi Ding & Tong Zhang (Tung Chang)

Elliptic and Parabolic Equations   30
     Guangcang Dong

Viscosity Solutions of Fully Nonlinear Elliptic and Parabolic
Equations   42
     Guangcang Dong & Baojun Bian

Some Developments of the Theory of Mixed PDEs   50
     Chaohao Gu & Jiaxing Hong

Free Boundary Problems   67
     Lishang Jiang

The Generalized Riemann Problem for Quasilinear Hyperbolic
Systems of Conservation Laws   80
     Ta-tsien Li

Minimal Surfaces in Riemannian Manifolds   104
     Guangyin Wang

Microlocal Analysis   111
     Rouhuai Wang & Shuxing Chen

Nonlinear Partial Differential Equations in Physics and Mechanics   127
     Yulin Zhou & Boling Guo

Part Two: Short Communications

Solitons and Exactly Solvable Nonlinear Evolution Equations   160
    Yi Cheng & Yi-shen Li

The Systems of Second Order Partial Differential Equations with
Constant Coefficients   173
    Wei Lin & Ciquian Wu

PREFACE

Before the founding of the People's Republic of China, there were only a few people who had researched partial differential equations. In 1936, during Professor Hadamard's visit to China, he delivered a series of lectures on the theory of second order linear partial differential equations. In 1951, Professor Sing-Mo Ou, who had studied ultra-hyperbolic equations under Professor Hadamard, returned to China and headed the division of differential equations at the Institute of Mathematics, Academia Sinica in 1953. Then in 1954, the Ministry of High Education of P.R.C. organized a summer school on partial differential equations and Professor Ou and Professor Huanwu Peng were each in charge of one course. Hundreds of young university or college teachers from all over China participated in the school and were very attracted by the lectures. Since then the subject of partial differential equations has been set up and research work on PDEs has spread to almost all universities in China. Meanwhile a number of young scholars were sent to the Soviet Union for further training. In 1956, among China's national scientific programmes, the subject of PDEs was placed as one of the most important subjects in mathematics, causing Chinese mathematicians to pay more attention to it. During that period, Professor Petrovski visited China and delivered lectures on PDEs. In 1957, Professor Bitsadze came to China to visit the Institute of Mathematics, Academia Sinica, and a number of young Chinese scholars gathered to listen to his lectures; some of them accomplished some research work under his guidance. After that the researchers of PDEs were directed to study some problems related to economic developments. Subjects included fluid mechanics, elastic mechanics, and other problems relating to nonlinear PDEs, such as the theory, the structure of solutions, and approximate analysis. A great deal of theoretical and applied work was done, and the level of research was raised significantly. Valuable results to some important problems were thus obtained.

However, in 1966 the Cultural Revolution began and many research programmes were severely impeded. The situation did not change until 1976. Since that time, China has opened her doors to the rest of the world. Academic research work received renewed attention and has recovered rapidly in China.

In the past few years China has invited many western scholars to share their know-how with their Chinese colleagues. They have also sent many bright young scholars or students abroad for advanced studies. As a result of this, scientific research in China has risen to a new state of vigour. In 1980 an International Conference on Differential Geometry and Differential Equations initiated by Professor S.S. Chern was held in Beijing, China. Many leading world-famous foreign mathematicians were invited and presented systematic lectures. Chinese mathematicians also shared their results with the visitors at the conference. This conference also played a role in summarizing the results of the past and shaped some future research trends. Since then international academic exchange has continued to develop rapidly. These activities have enabled Chinese scholars to broaden their methods and sharpen their research ability, and in turn, many foreign researchers have come to learn more about the achievements of Chinese scholars. However, we don't feel that the exchange is sufficient. For example, many results of Chinese scholars were published in Chinese Journals, in the Chinese language which are difficult to be studied or understood by western readers. The goal of this collection is to summarize and introduce the historical progress of the development of PDEs in China. As the research areas are very wide in PDEs the results covered in this book are mainly those that were published before the early 1980s.

We would like to express our thanks to those who contributed to the book in different ways; particularly we appreciate Professors Lishang Jiang, Ta-Tsien Li, and Tong Zhang greatly for their indispensable efforts and help rendered in the preparation of the book. We want to thank the publisher and mathematics editor of Kluwer Academic Publishers for their support and endorsement of the project.

Finally we want to express our deep gratitude to the late Professor Sing-Mo Ou for his great effort of promoting the research of PDEs in China and his endorsement during the early stage of the development of the collection.

Chaohao Gu   谷超豪

Xiaxi Ding   丁夏畦

Chung-Chun Yang   杨重骏

Contributors

Baojun Bian 边保军
  Department of Mathematics, Zhejiang University, Hangzhou, Zhejiang
  310027

Shuxing Chen 陈恕行
  Department of Mathematics, Fudan University, Shanghai 200433

Yazhe Chen 陈亚浙
  Department of Mathematics, Beijing University, Beijing 100871

Yi Cheng 程 艺
  Department of Mathematics University of Science and Technology of China,
  Hefei, Anhui 230026

Xiaxi Ding 丁夏畦
  Institute of Applied Mathematics, Academia Sinica, Beijing 100080

Guangcang Dong 董光昌
  Department of Mathematics, Zhejiang University, Hangzhou, Zhejiang 310027

Chaohao Gu 谷超豪
  Department of Mathematics, Fudan University, Shanghai 200433

Boling Guo 郭柏灵
  Centre for Nonlinear Studies Institute of Applied Physics and Computational
  Mathematics Beijing, P.O.Box 8009, Beijing 100088.

Jiaxing Hong 洪家兴
  Department of Mathematics, Fudan University, Shanghai 200433

Lishang Jiang 姜礼尚
  Department of Mathematics, Suzhou University, Suzhou, Jiangsu 215006

Ta-tsien Li 李大潜
  Department of Mathematics, Fudan University, Shanghai 200433

Yi-shen Li 李翊神
  Department of Mathematics University of Science and Technology of China,
  Hefei, Anhui 230026

Wei Lin 林 伟
  Department of Mathematics, Zhongshan University, Guangzhou, Guangdong
  510275

Guangyin Wang 王光寅
   Institute of Mathematics, Academia Sinica, Beijing 100080

Rouhuai Wang 王柔怀
   Department of Mathematics, Jilin University, Changchun, Jilin 130023

Ciquian Wu 吴滋潜
   Department of Mathematics, Zhongshan University, Guangzhou, Guangdong
   510275

Yulin Zhou 周毓麟
   Centre for Nonlinear Studies Institute of Applied Physics and Computational
   Mathematics Beijing, P.O.Box 8009, Beijing 100088

Tong Zhang 张 同
   Institute of Mathematics, Academia Sinica, Beijing 100080

Degenerate Parabolic and Elliptic Equations[†]

Yazhe Chen[††]

Department of Mathematics, Beijing University, Beijing

1 Linear and semilinear equations

The region where the equation detoriorates is fixed for linear and semilinear degenerate equations. The cases usually discussed are that the degenerate region is on the boundary. The two approaches are often used. One is the barrier argument and another is introducing the weighted Sobolev spaces.

1.1 The Keldysh type equations

Let Ω be a bounded domain on the upper half x-y plane with some part γ of the boundary $\partial\Omega$ on the x-axis. In 1950, M. V. Keldysh [KL] first studied the equation

$$y^{m_1} u_{xx} + y^{m_2} u_{yy} + a u_x + b u_y + c u = f \qquad (1.1)$$

in Ω where $m_1 \geq 0$, $m_2 \geq 0$, $m_1 + m_2 > 0$ and $m_1 \cdot m_2 = 0$. The equation (1.1) detoriorates on γ. We call γ the degenerate boundary of (1.1). Let $\Gamma = \partial\Omega \setminus \gamma$. He found that the Dirichlet problem for (1.1) is well-posed if

$$m_2 < 1; m_2 = 1, b|_\gamma < 1; 1 < m_2 < 2, b|_\gamma \leq 0; m_2 \geq 0, b|_\gamma < 0 \qquad (1.2)$$

and is not well-posed if

$$m_2 = 1, b|_\gamma \geq 1; 1 < m_2 < 2, b|_\gamma > 0; m_2 \geq 2, b|_\gamma \geq 0 \qquad (1.3)$$

† 1991 Mathematics Subject Classification: 35J70, 35K65, 35J25, 35J65, 35J60, 35K20, 35K15, 35K20, 35K55, 35K60, 35B65

†† Supported by NNSF of China

C. Gu et al. (eds.), *Partial Differential Equations in China*, 1–18.
© 1994 *Kluwer Academic Publishers*.

He further proved that if we impose a boundedness condition on the degenerate boundary γ instead of a Dirichlet condition the problem is still well - posed for the cases (1.3). S.A. Tersenov [TR] discussed the modified Dirichlet condition

$$u|_\Gamma = \phi, \qquad [\frac{u}{h}]_\gamma = \phi \tag{1.4}$$

for some properly chosen function h(x) in 1957, where ϕ is the Dirichlet datum. He proved that the problem (1.1), (1.4) is well-posed for the case $m_2 = 1$. Since 1958, Chinese mathematicians started a set of studies in Keldysh's equations. Z.Y. Hou [HZ] and G.J. Yang [YG] dealt with the problem (1.1), (1.4) for the cases $1 < m_2 < \frac{3}{2}$ and $\frac{3}{2} \leq m_2 < 2$ respectively. C.F. Wang [WC1] also investigated the modified Dirichlet problem (1.1), (1.4) for the case $1 \leq m_2 < 2$. G.C. Dong [DG1,2] discussed that the two cases (1.2), (1.3) occur on the degenerate boundary γ simultaneously.

O.A. Oleinik [OL] studied the oblique derivative problems for (1.1) in 1952. G.C. Dong [DG3] and L.J. Chen [CJ1] further studied the well-posedness of the problem with

$$u_y|_\gamma = \psi \tag{1.5}$$

or alternatively

$$[\frac{u_y}{h}]_\gamma = \psi \tag{1.6}$$

on the degenerate boundary γ instead of the Dirichlet condition. Dong found that under the conditions

$$a = O(y^s), \quad a_y = O(y^{s-1}), \quad s > \frac{m_2}{2} - 1$$

in the cases of $0 \leq m_2 < 1$ and $m_2 \geq 1$, $b|_\gamma < 0$ the problem (1.1), (1.5) is well-posed, and in the case $m_2 \geq 1$, $b|_\gamma \geq 0$ it is well posed only if we imposed the condition that $u_y|_\gamma$ is bounded instead of a given function. C.F. Wang [WC2] discussed the problem when the above two cases appear on γ at the same time for $m_2 = 1$.

For the corresponding semilinear equations

$$y^{m_1} u_{xx} + y^{m_2} u_{yy} + f(x, y, u, u_x, u_y) = 0 \tag{1.7}$$

Y.L. Zhou [ZY1] studied their first boundary value problems, where the growth order of f(x,y,u,p,q) with respect to $\sqrt{p^2 + q^2}$ is less than 2.

Because of the Cultural Revolution, the study broke up for thirteen years. In 1978, C.F. Wang [WC2] dealt with the well-posedness of the n-dimensional Dirichlet problems for the equations corresponding to (1.1) with $m_1 = m_2$. And then M.D. Li and Y.C. Qin [LQ] discussed the corresponding oblique derivative problems. Z.Z. Ding [DZ] studied the general boundary value problems for the semi-linear equations (1.1).

1.2   General degenerate equations

Let Ω be a bounded convex domain of R^n and Σ be the boundary of Ω. Consider the equation in divergence form

$$Lu = -D_i(a^{ij}D_j u) + b_i(x)D_i u + c(x)u = f(x) \tag{1.8}$$

where $a^{ij}(x) = a^{ji}(x)$ and

$$a^{ij}\xi_i\xi_j \geq \lambda(x)|\xi|^2, \quad \text{for} \quad any \quad x \in \Omega, \quad \xi \in I\!\!R^n \tag{1.9}$$

for some nonnegative function $\lambda(x)$. Here the summation convection is used. G. Fichera [FG] first divided the boundary Σ into four parts

$$\Sigma_0 = \{x \in \Sigma : a^{ij}n_i n_j = 0, b^i n_i = 0\},$$

$$\Sigma_1 = \{x \in \Sigma : a^{ij}n_i n_j = 0, b^i n_i > 0\},$$

$$\tag{1.10}$$

$$\Sigma_2 = \{x \in \Sigma : a^{ij}n_i n_j = 0, b^i n_i < 0\},$$

$$\Sigma_3 = \{x \in \Sigma : a^{ij}n_i n_j > 0\}$$

where $b^i(x)n_i$ is called the Frichera function. G. Fichera and O.A. Oleinik proved the existence and uniqueness of weak solutions if we impose the boundary value condition only on $\Sigma_2 \cup \Sigma_3$. They proved the existence, the uniqueness and the regularity of weak solutions of the above problems. Under the condition

$$\lambda^{-1}(x) \in L_s(\Omega), \qquad s > \frac{n}{2} \tag{1.11}$$

L.G. Gu [GL1], [GL2], [GL3] proved the existence and uniqueness of weak solutions in the energy space $\aleph$ and obtained the maximum principle of solutions only given the Dirichlet condition on Σ_3 using the Galerkin method and Moser iteration, where $\aleph$ is the completion $C^\infty(\overline{\Omega})$ with respect to the natural energy norm

$$||u||_\aleph = \{\int_\Omega [a^{ij}D_i u D_j u + u^2]dx\}^{\frac{1}{2}}$$

The space of rest function in the definition of weak solutions is denoted by $\aleph = \{u \in \aleph : u|_{\Sigma_3} = 0\}$. Then Gu and Zhuang [Gu4], [ZQ2] dealt with the corresponding eigenvalue problems and the Fredholm alternative theorems. After that the corresponding semilinear equations, the system with diagonal structure and the parabolic equations were studied by Gu and his group [GL5], [CZZ], [ZK1,2], [YH], [CZC1,2], [ZQ1]. Cao in [CZZ] studied the mixed boundary -value problem for the semilinear equation

$$D_j(a^{ij}(x)D_i u) + c(x)u = f(x, u) \quad \text{in} \quad \Omega$$

$$Bu = a^{ij}D_i u n_j + b(x)u = g(x, u) \quad \text{on} \quad \Sigma_3 \setminus \Gamma$$

$$u = 0 \quad \text{on} \quad \Gamma$$

By means of the critical point theory, he found non-trivial solutions to the above problem when f(x,u) and g(x,u) are sublinear or superlinear in u and discussed the problem about multiple solutions. Yang and Huang generalized the method in [FKS] to the degenerate elliptic system with diagonal structure

$$-D_\alpha\big(a^{\alpha\beta}(x)D_\beta u^i\big) + b^\alpha_{ij}(x)D_\alpha u^j + c_{ij}(x)u^j = f^i(x) - D_\alpha\big(g^\alpha_i(x) + d^\alpha_{ij}(x)u^j\big)$$

and obtained the Hölder continuity of weak solutions. Since the system is degenerate, the usual method of freezing the coefficients to obtain integral estimates is not suitable. They gave a different iteration approach to obtain the required integral estimates. They also discussed the quasilinear system

$$-D_\alpha\big(A^{\alpha\beta}(x,u)D_\beta u^i\big) + B_i(x,u,Du) = 0$$

where

$$|B_i(x,u,p)| \le \lambda\{b_1|p|^{1+\frac{\varsigma}{n}} + b_2|u|^{\frac{n+1}{n-1}} + f_i(x)\}$$

with $0 < \varsigma < 1$.

W.D., Lu [LW] established an embedding theorem for anisotropic Sobolev space and then found the existence of non-trivial solutions to the Dirichlet problem for a class of nonlinear anisotropic degenerate elliptic equations

$$-\sum_{i=1}^{n} D_i[\lambda_i(x)(D_i u)^{2m-1}] + c(x)u^{2m-1} = f(x,u) \quad \text{in} \quad Q_0$$

$$u = 0 \quad \text{on} \quad \partial Q_0$$

under some conditions on $f(x,\xi)$, where Q_0 is a cube in $\mathbb{R}^n$, using the embedding theorem and the mountain pass lemma.

Y.J. Shen [SY] applied the Moser iteration method to degenerate parabolic equations in divergence form, extending the results for elliptic equations in [FKS]. He first established the weighted imbedding theorem suitable for parabolic type. Under some conditions, He got the Harnack inequality for weak solutions and then Hölder continuity of weak solutions for uniformly degenerate parabolic equations and continuity for non-uniformly degenerate equations.

2   Quasilinear Equation

Generally the degenerate region for quasilinear equations depends upon the solution itself and so it is unknown. It cannot be solved by using the weighted Sobolev space as done in linear equations.

2.1 The equation deteriorating only on the boundary

Some quasilinear equations are degenerate only on the boundary if one imposed some restriction. The equations discussed are likely the generalization of linear Keldysh equations. Let Ω be a bounded domain in upper half space $R_+^n = R^n \cap \{x_n > 0\}$. The boundary consists of two parts: $\Lambda \subset R_+^n$ and $\gamma \subset \partial R_+^n$. Consider the quasilinear equation

$$a^{ij}(x, u)D_{ij}u + f(x, u, Du) = 0 \qquad (2.1)$$

Let $\lambda(x, u)$ and $\Lambda(x, u)$ be respectively the minimal and maximal eigenvalue of the coefficient matrix $[a^{ij}(x, u)]$ and assume

$$\lambda(x, 0) = 0; \qquad \lambda(x, u) > 0 \quad as \quad u \neq 0 \qquad (2.2)$$

which means the equation (2.1) is degenerate elliptic. We say that the equation (1.1) is of Keldysh type if

$$a^{nn}(x, 0) > 0 \quad as \quad x \in \overline{\Omega} \qquad (2.3)$$

or

$$a^{nn}(x, 0) = 0 \quad as \quad x \in \overline{\Omega}, \quad \exists k(1 \leq k \leq n-1)_{\ni},$$

$$a^{kk}(x, 0) > 0 \quad as \quad x \in \overline{\Omega} \qquad (2.4)$$

Consider the Dirichlet condition

$$u(x)|_\Gamma = \phi(x), \quad u(x)|_\gamma = 0 \qquad (2.5)$$

Before the Cultural Revolution, R.J. Shen [SR] studied problems of this type. Shen discussed the case that $a^{ij}(x, u)$ is degenerate like the power function u^m and the growth order of f(x,u,p) is less than 2 with repect to $|p|$. In the early eighties, L.S. Jiang [JL2], [JL3] dealt with two dimensional case with non-powerlike degenerate. He considered the equation

$$k(u)\frac{\partial^2 u}{\partial x^2} + \frac{\partial^2 u}{\partial y^2} + a(x, y)\frac{\partial u}{\partial x} + b(x, y)\frac{\partial u}{\partial y} + c(x, y)u = f(x, y) \qquad (2.6)$$

or

$$\frac{\partial^2 u}{\partial x^2} + k(u)\frac{\partial^2 u}{\partial y^2} + a(x, y)\frac{\partial u}{\partial x} + b(x, y)\frac{\partial u}{\partial y} + c(x, y)u = f(x, y) \qquad (2.7)$$

where $k(u) \in C[0, \infty)$, k(0)=0 and $k(u) > 0$ as $u > 0$. He proved the existence of a classical solution, which is positive in Ω, to the problem (2.6), (2.5) or (2.7), (2.5) under some conditions. His student J.L. Yue [YJL] generalized these results to the quasilinear equation (2.1) with $|f(x, u, p)| \leq C(|p|^2 + 1)$ in the cases (2.3) and (2.4). Moreover Yue also discussed uniformly degenerate case, i.e.

$$1 \leq \frac{\Lambda(x, u)}{\lambda(x, u)} \leq \nu \qquad (2.8)$$

The corresponding results for uniformly degenerate quasilinear elliptic equations in divergence form had been obtained [WLJY] by means of De Giorgi iteration in 1980. In order to guarantee the positiveness of solutions in the interior of Ω, some conditions on the terms of lower order must be imposed. L.S. Jiang and J.L. Yue [JY] discussed the influence of the terms of lower order.

2.2   General filtration equations

2.2.1 One-dimensional problems

Y.L. Zhou [OKZ] first make initiative and systematic researches into general filtration equations with Oleinik and Kalashnikow during his study at Moscow University. Their work is well-known in this field. After the Cultural Revolution, many Chinese mathematicians started to engage in research into these problems. In view of the needs of national economy. the flow of underground water through porous media was studied extensively. It is governed by the porous medium equation with a gravitation term

$$u_t = (a(u)u_x)_x + b(u)u_x \tag{2.9}$$

where $a(s) \in C[0, \infty)$, a(0)=0 and $a(s) > 0$ as $s > 0$. There is a gap between the existence conditions and the uniqueness conditions when Gilding [GI] studied the equation (2.9). For the existence of nonnegative weak solutions Gilding supposed that

$$sb'(s) \in L^1(0, 1) \tag{2.10}$$

but for the uniqueness he set a rather strict restriction

$$b^2(s) = O(a(s)) \quad \text{as } s \to +0 \tag{2.11}$$

D.Q. Wu [WD1] and G.C. Dong, Q.X. Ye [DY] improved the uniqueness condition (2.11), but they still required some relations of b(s) to a(s). Y.Z. Chen finally removed these relations. In the above works on uniqueness it is required that a(s) is power-like near s=0. Using the L^1 estimation technique, J.L. Zhao [ZJ2] obtained a very good uniqueness result. He consider the quasilinear degenerate parabolic equation

$$u_t = (a(x, t, u)u_x)_x + (b(x, t, u))_x + c(x, t)u + f(x, t) \tag{2.12}$$

where $a(x, t, u) \geq 0$. The only structure condition for uniqueness of bounded measurable solutions is that

$$A(x, t, u) = \int_0^u a(x, t, s)ds \tag{2.13}$$

is strictly increasing with respect to u. S.Q. Wang [WS] generalized the other results in [GI] and established the equation satified by the interface between dry and wet states. L.S. Jiang [JL4] found the better regularity of weak solutions in the case of $a(s) = ms^{m-1}$. He obtained that for $0 \le \alpha < 1$, $A_\alpha(u) = u^{m-1+\alpha} \in C^{1,\frac{1}{2}}(\overline{Q}_T)$ or $u \in C^{\lambda,\frac{\lambda}{2}}(\overline{Q}_T)$ with $\lambda = min\{(m-1+\alpha)^{-1}, 1\}$ if $|s^\alpha b(s)|$ and $|s^{1+\alpha}b'(s)|$ are bounded in $(0,\infty)$. D.Q. Wu [WD2] discussed the equation (2.12) with singular coefficients

$$a(x,t,u) = O(u^\gamma), \quad b(x,t,u) = O(u^\mu), \quad -1 < \gamma, \mu < 0$$

The problems mentioned above are all about the filtration in insaturated porous media. Ones also consider the filtration in partial saturated porous media. It is governed by

$$c(u)_t = u_{xx} - K(u)_x \tag{2.14}$$

or

$$c(u)_t = u_{xx} \tag{2.15}$$

if the gravitational effect is neglected, where

$$c(s) \in C^{0,1}(\mathbb{R}), \quad c'(s) > 0 \quad \text{as} \quad s < 0 \quad \text{and}$$
$$c(s) = 1 \quad \text{as} \quad s > 0 \tag{2.16}$$

Van Duijn and Peletier [VD], [VP] obtained global existence, uniqueness and asymptotic behavior of weak solutions and dealt with the interface between unsaturated and saturated states for the equation (2.15). L.J. An [AL1], L. Su and J.G. Li [SL], [LS] investigated the Cauchy problem, the first and second boundary-value problems for (2.14). The corresponding results to [VD] and [VP] were established. The degenerate of two kinds (dry state or saturated state) were studied separately in the above works. S.T. Xiao and his group [XHZ], [AL2] first studied the situation in which the degeneration of two kinds appears at the same time. The equation is still (2.14), but $c(s)$ satisfies

$$\begin{cases} c(s) \in C^{0,1}[0,\infty) \quad \text{is} \quad \text{strictly increasing} \quad \text{on} \quad [0,\infty), \\ c(0)=0 \quad \text{and} \quad c(s)=1 \quad \text{as} \quad s \ge 1 \end{cases} \tag{2.17}$$

The problem with constant surface flux

$$\begin{cases} c(u)_t = (u_x - K(u))_x \quad \text{in} \quad Q_T = (0,\infty) \times (0,T] \\ (u_x - K(u))|_{x=0} = -R \quad \text{on} \quad [0,T] \\ crc(u(x,0)) = v_0(x) \quad \text{on} \quad [0,\infty) \end{cases} \tag{2.18}$$

was discussed. They got the existence, uniqueness and asymptotic behavior of weak solution and study when the water content $c(u(x,0))$ has compact support and then established the existence and regularity of the interfaces of two kinds.

Y.L. Zhou [ZY2] studied the degenerate parabolic system of filtration type

$$u_t = (grad\phi(u))_{xx} \tag{2.19}$$

where $u = (u_1, u_2, ..., u_N)$ is a vector value function and $\phi(u)$ is a strictly convex function with Hessian matrix zero-definite at u=0. Using the regularization method, he established the existence and uniqueness of generalized solutions to the periodic boundary - value problem and the Cauchy problem for (2.19). H.Y. Fu [FH] consider the more general system

$$u_t = f(u)_{xx} \tag{2.20}$$

where the vector value function f(u) satisfies the monotonic condition and

$$\exists \alpha, \beta > 0_\ni \quad < u, f(u) > \geq \alpha|u|^{1+\beta}$$

He proved that the difference solution converges to the weak solution. The uniqueness was also got.

2.2.2 multi-dimensional problems

Y.Z. Chen studied the general filtration equation

$$u_t = D_i(a^{ij}(x,t,u)D_j u) + b^i(x,t,u)D_j u + c(x,t,u)u \tag{2.21}$$

in $Q_T = \Omega \times (0,T] \subset \mathbb{R}^{n+1}$, where a^{ij} satisfies

$$\nu(|u|)|\xi|^2 \leq a^{ij}(x,t,u)\xi_i\xi_j \leq \Lambda\nu(|u|)|\xi|^2, \quad \forall(x,t,u) \in Q_T \times \mathbb{R}, \ \xi \in \mathbb{R}^n \tag{2.22}$$

$$\nu(s) \in C[0,\infty), \quad \nu(0) = 0 \quad \text{and} \quad \nu(s) > 0 \quad \text{as} \quad s > 0 \tag{2.23}$$

and $\nu(s)$ is a power-like function, i.e. $\nu(s)$ satisfies

$$1 \leq \frac{s\phi'(s)}{\phi(s)} \leq M \tag{2.24}$$

near s=0, where $\phi(s) = \int_0^s \nu(\tau)d\tau$. Using De Giorgi iteration with some scaling factor in time for the iteration boxes, he [CY2] first established the Hölder estimates for weak solutions to (2.21) under an additional structure condition

$$\sum_{i=1}^n [b^i(x,t,u)]^2 \leq \Lambda\nu(|u|) \tag{2.25}$$

Consequently, the existence and Hölder continuity of weak solutions to the Cauchy problem and the first boundary-value problem for (2.21) were obtained easily and the uniqueness was also established [CY3]. G.C. Dong [DG5] improved the proof and removed the additional structure condition (2.25). Developing the technique of De Giorgi iteration. Y.Z, Chen [CY4] also gave a global description of finte propogation speed for the equation (2.21). Instead of (2.23) assume that $\nu(s)$ satisfies

$$m^{-1}(\frac{s_1}{s_2})^{1-\frac{1}{m}} \leq \frac{(\phi^{-1})'(s_2)}{(\phi^{-1})'(s_1)} \leq \lambda(\frac{s_1}{s_2}) \tag{2.26}$$

for $0 < s_1 < s_2$, where $\lambda(\tau)$ is non-decreasing and $\lambda(\tau) \to 0$ as $\tau \to 0^+$. The weak solution $u(x,t)$ to (2.21) satisfies that, for any $(x^0, t^0) \in Q_T = \mathbb{R}^n \times (0, T]$,

$$C^{-1} min\{ \inf_{|x-x^0| \le b\sqrt{t^0}} u(x, 0), 1 \} \le u(x^0, t^0) \le C \sup_{|x-x^0| \le b\sqrt{t^0}} u(x, 0) \qquad (2.27)$$

if $|u(x, t)| \le M$, where the constants b and C depend only upon M, T and the coefficients of (2.21).

Q.B. Huang [HQ] generalized the existence and uniqueness results in [WJ] to the equation

$$u_t = \Delta\phi u + div B(u) + F(u)$$

in some singular cases. He also studied blow-up of weak solutions of the equation

$$[\beta(u)]_t = \Delta u + F(u) \quad in \quad Q_T = \Omega \times (0, T]$$

under the initial and boundary value conditions

$$u = 0 \quad on \quad \partial\Omega \times (0, T), \qquad u(x, 0) = u_0(x)$$

where $\beta(s) = k_1(s)s^m$, $F(s) = k_2(s)\beta(s)^{1+\sigma}$ ($s > 0$) with $m \ge 1$, $\sigma > 0$, $k_1(0)$, $k_2(0) > 0$, and k_1, k_2 increasing. Denote

$$H(0) = \int_\Omega \int_0^{u_0} \beta'(s)s\,ds\,dx$$

He proved under some smoothness conditions that if $u_0 \in L^\infty(\Omega) \cap H_0^1(\Omega)$, $u_0 \ge 0$ and

$$\frac{m+1}{m\sigma c_0} H(0)^{-m\sigma/(m+1)} \le T, \quad 2\int_\Omega \int_0^{u_0} F(s)\,ds\,dx - \int_\Omega |\nabla u_0|^2\,dx \ge 0,$$

with some cnstant c_0, then the solution $u(x,t)$ must blow up by T.

S.Y. Huang [FH] studied the mesa problem with A. Friedman. They considered a sequence of evolutionary problems

$$\begin{cases} u_t = \Delta\phi_m(u) & in \ \mathbb{R}^n \times (0, \infty) \\ u(x, 0) = f(x) & in \ \mathbb{R}^n \end{cases} \qquad (2.28)$$

and assumed that

$$\phi_m(r) \text{ is continuous and nondecreasing in } [0,\infty);$$
$$\phi_m(0) = 0, \phi_m(r) \to 0 \text{ as } m \to \infty \text{ for } 0 < r < 1$$
$$\phi_m(r) \to \infty \text{ as } m \to \infty \text{ for } 1 < r < \infty$$

and that

$$f \in L^1(\mathbb{R}^n) \cap L^\infty(\mathbb{R}^n), \quad f \ge 0$$

the set $\{x \in \mathbb{R}^n | f(x) > 1\}$ has positive measure.

Obviously $\phi_m(r) = r^m$ is an example. For each m, there exists a unique nonnegative weak solution $u^{(m)}$ to the problem (2.28). They proved that $u^{(m)}$ is weakly star convergent to the function

$$u^{(\infty)} = 1 \cdot \chi_A + f(x)\chi_{\mathbb{R}^n \setminus A}$$

in $L^\infty(\mathbb{R}^n \times (0, \infty))$, where the set A is the noncoincidence set of the solution of the variational inequality

$$-\Delta w \geq f(x) - 1, \quad w \geq 0, \quad w(\Delta w + f - 1) = 0 \quad \text{in} \quad \mathbb{R}^n$$

The typical degenerate parabolic equation of another kind is the evolutionary p-Lapacian equation

$$u_t - div(|\Delta u|^{p-2}\Delta u) = 0, \qquad 1 < p < \infty \tag{2.29}$$

and the corresponding system

$$u_t - div(|\Delta u|^{p-2}\Delta u^\alpha) = 0, \quad \alpha = 1, 2, ..., m. \quad 1 < p < \infty \tag{2.30}$$

which describe the flow of non-Newtonian fluid. Y.T. Shen and X.K. Guo [SY], [GS] discussed the equation

$$u_t - D_i(a^i(x, t, u, u_x)) + b(x, t, u, u_x) = 0 \tag{2.31}$$

where a^i and b satisfy

$$\int_\Omega [a^i(x, t, u, u_x)D_i u + b(x, t, u, u_x)u]dx$$

$$\geq \lambda \int_\Omega |u_x|^p dx - C \int_\Omega (1 + u^2)dx, \quad \lambda > 0$$

and

$$\int_\Omega [a^i(x, t, v, v_x) - a^i(x, t, v, u_x)](D_i v - D_i u)dx$$

$$\geq \int_\Omega \mu(|v|, |u_x|)|v_x - u_x|^2 dx, \quad \mu(., .) > 0$$

for any $v, u \in W_0^{1,p}(\Omega)$. By using Galerkin methods they proved the existence of weak solutions under some other structure conditions. L.K. Gu and Z.C. Cao [CZC1], [GC] dealt with the equation (2.31) with the following degenerate structure

$$\frac{\partial a^i(x, t, u, p)}{\partial p_j}\xi_i\xi_j \geq \lambda \sum_k |a^{kj}(x)p_j|^{p-2}|\sum_k a^{kj}(x)\xi_i|^2$$

where the matrix $(a^{kj}(x))$ is degenerate. Denote $\Sigma_3 = \{x \in \partial\Omega | a^{kj}(x)n_j \neq 0, k = 1, 2, ..., n\}$ like (1.10). Under some structure condition they obtained the existence and uniqueness of weak solutions to the equation (2.31) with the initial-boundary value conditions

$$u(x, 0) = u_0(x), \quad x \in \Omega, \qquad u(x, t) = 0 \quad (x, t) \in \Sigma_3 \times (0, T]$$

The regularity of weak solutions to (2.29), (2.30) or more general equations is an very interesting problem. Combining the technique of scaling a factor in the iteration boxes with the method in [EV] and [DF], Y.Z. Chen found the local Hölder continuity of the gradient of weak solution to the equation (2.29) and the system (2.30) for the degenerate case $p > 2$ independently. X.F. Chen [CX] generalized it to more general equations. Y.Z. Chen studied the boundary regularity of weak solutions to the Dirichlet problem for the system (2.30) with E. Di Benedetto. They obtained the global Hölder continuity of gradient if the Dirichlet datum is zero, and C^α regularity of weak solutions for any $0 < \alpha < 1$ for general Dirichlet data under some assumptions.

For the equation

$$u_t - D_j\big(a^{ij}(x,t)|\nabla u|^{p-2}D_i u\big) = 0, \quad 1 < p < \infty \tag{2.32}$$

with (a^{ij}) bounded measurable and

$$\lambda|\xi|^2 \le a^{ij}\xi_i\xi_j \le \Lambda|\xi|^2, \quad \text{for any } \xi \in I\!\!R^n, \quad \lambda > 0$$

E. Di Benedetto [DB] established Hölder continuity of weak solutions to the equation (2.32) or much more general equations with the corresponding structure for $p > 2$. The problem for the singular case $1 < p < 2$ is much more difficult. Y.Z. Chen and Di Benedetto [CD] solved the problem for a class of quasilinear equations with $1 < p < 2$.

2.3 Strongly degenerate equations

Vol'pert and Hujaev [VH] first devoted to the global solutions of the quasilinear strongly degenerate parabolic equation

$$Lu = u_t - D_i\big(a^{ij}(x,t,u)D_j u\big) - D_i\big(f(x,t,u)\big) = g(x,t,u) \tag{2.33}$$

with $a^{ij} = a^{ji}$ and
$$a^{ij}(x,t,u)\xi_i\xi_j \ge 0, \qquad \forall \xi \in I\!\!R^n$$

Since (a^{ij}) is strongly degenerate, solutions may be discontinuous. Vol'pert and Hujaev established the existence and uniqueness of global solutions in BV space of the Cauchy problem for the equation (2.33) in one space variable. Z.Q. Wu and J.Y. Wang [WW] generalized it to the first boundary value problem. Since the equation may be degenerate on the part of the boundary, the problem has some essential difficulty. Then Z.Q. Wu and J.L. Zhao [WZ] discussed the problem in several space variables. They assumed that (a^{ij}) also satisfies

$$a^{ij}\xi_i\xi_j - \delta \sum_{s=1}^{n+1} \sum_{j=1}^{n} (a_{x_s}^{ij}\xi_i)^2 \ge 0 \quad \forall \xi \in I\!\!R^n$$

$(x_{n+1} = t)$ for some positive constant δ. By means of the method of parabolic regularization, they proved the existence of BV generalized solutions to the first boundary value problem for (2.33). The most difficult point is the L^1 estimate for the gradient of the approximating solutions. However, the uniqueness was established only for a narrow class of BV solutions, which satisfy

$$r^{ij}(x, t, u) D_i u \text{ is bounded and is in } BV(\Omega) \text{ in } x$$

where (r^{ij}) is the square root of the matrix (a^{ij}). In [WY], Z.Q. Wu and J.X. Yin solved completely the uniqueness of BV generalized solutions in one space variable. Vol'pert and Hujaev [VH] had tried to prove a similar result, but their proof is incorrect due to the adoption of the wrong form of discontinuity condition and then proved the uniqueness of BV generalized solutions to the Cauchy problem in one space variable. The proof is based on the study of $BV_x(Q_T)$ which has independent interest. J.L. Zhao [ZJ4] also studied the properties of BV solutions. He proved that if (x_0, t_0) is a point of approximate continuity of the BV solution u(x,t) such that

$$a^{ij}(x_0, t_0, u(x_0, t_0)) \xi_i \xi_j > 0, \quad \forall \xi \in \mathbb{R}^n, \quad \xi \neq 0$$

then u(x,t) is a classical solution in the neighborhood of (x_0, t_0). In addition, he found that if A(x,t,u) is strictly increasing in u then BV solution is continuous for one space variable. Applying the theory of compensated compactness, J.L. Zhao discussed the existence of L^∞ generalized solutions to the first boundary value problem for (2.33) with

$$a(x, t, u)|\xi|^2 \leq a^{ij}(x, t, u) \xi_i \xi_j \leq \Lambda a(x, t, u)|\xi|^2, \quad \forall \xi \in \mathbb{R}^n$$

$$(x, t, u) \geq 0$$

He proved that if for almost $(x, t) \in Q_T$, A(x,t,u) defined in (2.13) is strictly increasing in u then the first boundary value problem for (2.33) has at least a L^∞ generalized solution under smoothness conditions.

2.4   Higher order equations

J.X. Yin [YJX2,3] studied the equation of fouth order in one space variable

$$u_t + D^4 A(u) = f \tag{2.34}$$

The basic assumptions are

$$A'(s) \geq 0 \qquad \forall s \in \mathbb{R}$$

$$\lim_{s \to +\infty} A(s) = +\infty, \qquad \lim_{s \to -\infty} A(s) = -\infty$$

He obtained the existence of L^∞ generalized solutions to the first boundary value problem for (2.34) and the uniqueness of generalized solutions in the class $\{u \in$

$L^2(Q_T)|A(u) \in L^2(Q_T)\}$. He found that if $A(s)$ is strictly increasing the generalized solutions are continuous and furthermore if $A(s)$ also satisfies

$$|A(s_1) - A(s_2)| \geq \alpha_0 |s_1 - s_2|^m \quad \forall s_1, s_2 \in \mathbb{R}$$

for some $\alpha_0 > 0$, $m > 0$ the generalized solutions are Hölder continuous. He [YJX1] also discussed the existence of classical solutions to the eqaution

$$u_t + D^4 u^{2m+1} + \lambda u = f \quad (\lambda \geq 0, m \geq 7)$$

and found that nonnegative classical solutions are not diffuse as the time t increases if $f = 0$.

References

[Al1] An Lian-Jun, Some properties of the solutions of a filtration problem in partial saturated porous media, Acta Math. Appl. Sinica, 1(1984), No.1, 44-56.

[AL2] An Lian-Jun, The infiltration problem with large constant surface flux in partially saturated porous media, Acta Math. Appl. Sinica, 2(1985), 332-345.

[BD] Bai Dong-Hua, Unsteady flow of power law fluids in porous medium with double porosity, J. PDE, 2, No. 4(1989), 67-82.

[BL] Bai Dong-Hua & Liu Yi, Parabolic systems without quasi-monotony , J. PDE, 2, No.3(1989), 62-78.

[CZC1] Cao Zhen-Chao, The boundary value problem for a class of quasilinear degenerate elliptic equations, Acta Sci. Natr. Univ. Amoien, 25 (1986), 384-399 (in Chinese)

[CZC2] Cao Zhen-Chao, Counter examples of the results on the mixed boundary value problem for second order quasilinear degenerate parabolic equations, Acta Sci. Natr. Univ. Amo. 24 (1985), 142-146 (in Chinese).

[CZC3] Cao Zhen-Chao, The Cauchy problem for degenerate parabolic equations, Acta Sci. Natr. Univ. Amoien, 24 (1985), 424-431 (in Chinese).

[CZZ] Cao Zhen-Zong, Non-zero solutions of the boundary value problem for nonlinear degenerate elliptic equations, 27 (1988), 159-164 (in Chinese).

[CC] Chen Chang-Sheng, The propagation of nonstationary filtration with absorption, Acta Math. Appl. Sinica, 3(1987), 342-350.

[CL] Chen Liang-Jing, A boundary value problem for degenerate elliptic equations, Acta Math. Sinica, 13(1963), 332-342.

[CX] Chen Xin-Fu, Hölder estimate for the solution of degenerate parabolic equation, Acta Math. Appl. Sinica, 13 (1987), No. 1, 70-96.

[CY1] Chen Ya-Zhe, Uniqueness of weak solutions of quasilinear parabolic degenerate equations, The Proc. of the 1982 Changchun Sym. on Diff. Geom. and Diff. Eqns., 317-332.

[CY2] Chen Ya-Zhe, Hölder estimates for solutions of uniformly degenerate quasilinear parabolic equations, The Proc. of the 1982 Changchun Sym. on Diff. Geom. and Diff. Eqns., 333-336 (Abstract), Chin. Ann. Math., 5(1984), 661-678.

[CY3] Chen Ya-Zhe, Existence and uniqueness of weak solutions of uniformly degenerate quasilinear parabolic equations, Chin. Ann. Math., 6 (1985), 131-146.

[CY4] Chen Ya-Zhe, On finite diffusing speed for uniformly degenerate quasilinear parabolic equations, Chin. Ann. Math., 7(1986), 318-329.

[CY5] Chen Ya-Zhe, Hölder continuity of the gradient of the solutions of certain degenerate parabolic equations, Chin. Ann. Math., 8(1987), 343-356.

[CY6] Chen Ya-Zhe, Hölder continuity of the gradient of solutions of nonlinear degenerate parabolic systems, Acta Math. Sinica (English Issue), 2 (1986), 309-311.

[CD1] Chen Ya-Zhe & Di Benedetto E., On the local behavior of solutions of singular parabolic equations, Arch. Rat. Mech. Anal., 103(1988), 319-345.

[CD2] Chen ya-Zhe & Di Benedetto E., Boundary estimates for solutions of singular parabolic systems, Jour. fur die Reine und Angew. Math., 395(1989), 102-131.

[CQ] Chen Zu-Chi & Qian Chun-Lin, A class of boundary value problems for quasilinear degenerate elliptic equations in permeating problems, The Proc. of the 1982 Changchun Sym. on Diff. Geom. and Diff. Eqns., 337-352.

[DB] E. Di Benedetto, On the local behavior of solutions of degenerate parabolic equations with measurable coefficients, Ann. Sc. Norm. Sup. Pisa Cp. Ser. IV, 8(1986), 487-535.

[DF] E. Di Benedetto & A. Friedman, Regularity of solutions of nonlinear degenerate parabolic systems, J. Reine Angew. Math., 349(1984), 83-128.

[DZ] Ding Zheng-Zhomg, The general boundary problem for a class of semilinear degenerate elliptic equations of second order, Acta Math. Sinica, 27(1984), 177-191.

[DG1] Dong Guang-Chang, A boundary-Value problem for degenerate elliptic equations, Acta Math. Sinica, 11(1961), 371-375 (in Chinese).

[DG2] Dong Guang-Chang, A correction to the paper " A boundary - value problem for degenerate elliptic equations", Acta Math. Sinica, 13(1963) (in Chinese).

[DG3] Dong Guang-Chang, Boundary-value problems for degenerate elliptic differential equations, Acta Math. Sinica, 13(1963), 94-115 (in Chinese).

[DG4] Dong Guang-Chang, A boundary value problem for the degenerate elliptic partial differential equation, Sc. Sinica, 13(1964), 697-708.

[DG5] Dong Guang-Chang, The first boundary value problem for solutions of degenerate quasilinear parabolic equations, Chin. Ann. Math., 7B(1986), 277-302.

[DG6] Dong Guang-Chang, A higher dimensional nonlinear degenerate parabolic equation, The Proc. of the 1982 Changchun Sym. on Diff. Geom. and Diff. Eqns., 369-375.

[DY] Dong Guang-Chang & Ye Qi-Xiao, On the uniqueness of solutions of nonlinear degenerate parabolic equations, Chin. Ann. Math., 3(1982), 279–284.

[EV] C.L. Evans, A new proof of local $C^{1+\alpha}$ regularity for solution of certain degenerate elliptic P.D.E., J. Diff. Eqns., 45(1982), 365-373.

[FKS] E.B. Fabes, C.E. Kenig & R.P. Serapioni, The local regularity of solutions of degenerate elliptic equations, Comm. in PDE, 7(1982), 77- 116.

[FHS] A. Friedman & Huang Shao-Yun, Asymptotic behavior of solutions of $u_t = \Delta\phi_m(u)$ as m → ∞ with inconsistent initial values, Analyse Math. Appl., Gauthier-Villars, Paris, 1988, 165-180.

[FH] Fu Hong-Yuan, Initial and boundary problems for the degenerate or singular system of the filtration type, Lecture Notes in Math., 1306, 69-83.

[GI] B.H. Gilding, A nonlinear degenerate parabolic equation, Annali della Scuola Norm. Sup. di Pisa, 4(1977), No. 3, 393-432.

[GL1] Gu Lian-Kun, The maximum principle of degenerate ellptic equations, Sc. Tong-bao, No.12(1981), 763(in Chinese).

[GL2] Gu Lian-Kun, Galerkin method in the boundary value problem for degenerate elliptic equations, Sc. Tongbao, No. 1(1982), 62(in Chinese).

[GL3] Gu Lian-Kun, The boundary value problem for degenerate elliptic equations, Acta Math. Sinica, 27(1984), 69-81(in Chinese).

[GL4] Gu Lian-Kun, The a note on the eigenvalue problem for degenerate elliptic equations, Adv. in Math., 13(1984), 318-320(in Chinese).

[GL5] Gu Lian-Kun, The boundary value problem for semilinear degenerate elliptic equations, Sc. Tongbao, No. 19(1982), 1213(in Chinese).

[GC] Gu Lian-Kun & Cao Zhen-Zhao, The initial-boundary value problem for a class of quasilinear parabolic equations, Acta Sci. Natr. Univ. Amoien, 26(1987), 1-10(in Chinese).

[GS] Guo Xin-Kang & Shen Yao-Tian, On the first boundary value problem for quasilinear parabolic equation with strongly singular coefficients, The Proc. of the 1982 Changchun Sym. on Diff. Gepm. and Diff. Eqns., 437-444.

[HZ] Hou Zong-Yi, Dirichlet problems for a class of second order linear elliptic equations deteriorating on the boundary of the domain, Science Records, New Ser., 2(1958), 311-315.

[HQ] Huang Qing-Bo, The boundary value problem for a class of degenerate and singular nonlinear diffusion equations, Northeastern Math. J., 5(1989), 155-169(in Chinese).

[HZ] Huang Zhi-Da, The generalized solutions and asymptotic solutions of infiltration problems with constant flux, Acta Math. Sinica, 26(1983), 677 -697(in Chinese).

[JL1] Jiang Li-Shang, Exact solutions of filtration systems in double porous media, Scien. Sinica, 1980, 152-165(in Chinese).

[JL2] Jiang Li-Shang, Quasilinear degenerate elliptic differential equations, Chin. Ann. MAth. (English issue), 2(1981), 41-52.

[JL3] Jiang Li-Shang, Remarks on quasilinear degenerate elliptic equations, Proceeding of the Beijing Sym. on Diff. Geom. and Diff. Eqns., 1980, 1273-1276.

[JL4] Jiang Li-Shang, Regularity and uniqueness of a weak solution for a quasilinear degenerate parabolic equation, Acta Math. Sinica, 29(1986), 1-9.

[JY] Jiang Li-Shang & Yue Jiang-Liang, Influences of the first derivative terms on solutions for a class of quasilinear degenerate elliptic equations, J. of Peking Univ. Natur. Sc., 1985, No.5, 1-7(in Chinese).

[KL] Keldysh M.V., On some cases of degenerate equations of elliptic type on the boundary of domains, Dokl. Acad. Nauk. USSR, 77(1951), 181-183.

[LS] Li Jian-Guo & Su Ling, The solution of an infiltation problem with ponded surface flux condition, Acta Math. Appl. Sinica, 2(1985), No.1, 54-65.

[LQ] Li Ming-De & Qin Yu-Chun, The oblique derivative problem for a class of sinular elliptic partial differential equations, Hangzhou Univ. Natur. Sc., 2(1980), 1-9.

[LW1] Lu Wen-Duan, Imbedding theorem of spaces of anisotropic integrable functions and Dirichlet problem for a class of degenerate elliptic equations, Proc. of the 1982 Changchun Sym. on Diff. Geom. and Diff. Eqns., 523-527.

[LW2] Lu Wen-Duan, An imbedding theorem for spaces of anisotropic integrable functions, Acta Math. Sinica, 27(1984), 319-344.

[OL]   O.A. Oleinik, On elliptic equations degenerate on the boundary of boundary, Dokl. Acad. Nauk. SSSR, 87(1952), 6, 885-888.

[OKZ] Oleinik O.A., Kalashnikow A.S. & Zhou Yu-Lin, The Cauchy problem and boundary problems for equations of the type of nonstationary filtration, Isv. Akad. Nauk. SSSR, Ser. Mat., 22(1958), 667-704.

[SR]   Shen Rong-Jun, Boundary value problems for nonlinear degenerate elliptic equations, Zhejiang Univ. Natur. Sc., 3(1965), 1-36.

[SYT] Shen Yao-Tian, On the Dirichlet problem for quasilinear elliptic equation with strongly singular coefficients, The Proc. of the 1980 Beijing Sym. on Diff. Geom. and Diff. Eqns., Vol.3, 1407-1417.

[SYJ] Shen You-Jian, A Harnack inequality for degenerate parabolic equations, J. PDE, 2, No.4(1989), 1-21.

[SL]   Su Ling & Li Jian-Guo, The boundary value problems for the filtration equation in partially saturated porous media, Acta Math. Appl. Sinica, 1(1984), 180-192.

[VD]   C.J. Van Duijn, Nonstationary filtration in partially staturated porous media: continuity of the free boundary, Arch. Rat. Mech. Anal., 79 (1982), No.3, 261-265.

[VP]   C.J. Van Duijn & L.A. Peletier, Nonstationary filtration in partially saturated porous media, Arch. Rat. Mech. Anal., 78(1982), No.3, 173-198.

[VY]   Van Duijn C.J. & Ye Qi-Xiao, The flow of two immiscible fluids through a porous medium, Nonlinear Anal., 5(1984).

[VH]   , A.I. Vol'pert & S.I. Hujaev, Cauchy's problem for second order quasilinear degenerate parabolic equations, Mat. Sb., 78(120)(1969), 389- 411.

[WC1] Wang Chuang-Fang, A boundary problem for a class of degenerate elliptic equations, Hangzhou Univ. Xuebao, 1(1963), 37-46,(in Chinese).

[WC2] Wang Chuang-Fang, A boundary value problem for degenerate elliptic PDEs, Hangzhou Univ. Xuebao, 2(1963), 31-38(in Chinese).

[WC3] Wang Chuang-Fang, Dirichlet problems for singular elliptic partial differential equations, Hangzhou Univ. Natur. Sc., 2(1978), 19-32 (in Chinese).

[WS]   Wu Shu-Qiao, The lateral boundary of the support of weak solutions of the filtration equation in one dimension, Acta Math. Sinica, 26(1983), 199 -219.

[WD1] Wu De-Quan, Uniqueness of the weak solution of quasilinear degenerate parabolic equations, Porc. of the Beijing Sym. on Diff. Geom. and Diff. Eqns., 1980 (Abstract), Acta Math. Sinica, 25(1982).

[WD2] Wu De-Quan, Existence of continuous solutions of quasilinear degenerate parabolic equations with singular coefficients, J. PDE, 1B, No.2 (1988), 77-83(in Chinese).

[WJ]   Wu De-Quan & Jiang Li-Shang, On the boundary value problem of a multidimensional quasilinear parabolic equations, Northeastern Math. J., 1(1985), 54-67(in Chinese).

[WLJY] , Wu Lang-Cheng, Liu Xi-Yuan, Jiang Li-Shang & Yue Jing-Liang, On Dirichlet problem of uniformly degenerate quasilinear elliptic equations, Proc. of the Beijing Sym. on Diff. Geom. and Diff. Eqns., 1980, 1581-1592.

[WW]   Wu Zhuo-Qun & Wang Jun-Yu, Some results on quasilinear degenerate parabolic equations of second order, Proc. of the Beijing Sym. on Diff. Geom. and Diff. Eqns., 1980.

[WY] Wu Zhuo-Qun & Yin Jing-Xue, Some properties of functions in BV_x and their applications to the uniqueness of solutions for degenerate quasilinear parabolic equations, Northeastern Math. J., 5(1989), 395-422.

[WZ1] Wu Zhuo-Qun & Zhao Jun-Ning, First boundary value problem for quasilinear degenerate parabolic equations of second order in several space variables, Proc. of the 1982 Changchun Sym. on Diff. Geom. and Diff. Eqns., 659-662.

[WZ2] Wu Zhuo-Qun & Zhao Jun-Ning, The first boundary value problem for quasilinear parabolic equations of second order in several space variables, Chin. Ann. Math., 4B(1983), 87-96.

[XHZ] Xiao Shu-Tie, Huang Zhi-Da & Zhou Chuan-Zhong, The infiltration problem with constant rate in partially saturated porous media, Acta Math. Appl. Sinica, 1(1984), 108-126.

[YG] Yang Guang-Jun, On Dirichlet problems for a class of degenerate elliptic equations, Acta Math. Sinica, 12(1962), 40-46(in Chinese).

[YS] Yang Shi-Xin, The boundary value problem for a class of quasilinear degenerate elliptic equations, Acta Sci. Natr. Univ. Amoien, 24(1985), 413-423(in Chinese).

[YH] Yang Shi-Xin & Huang Qin-Bo, The local regularity of weak solutions of diagonal type degenerate elliptic systems, Acta Sci. Natr. Univ. Amoien, 27(1988), 20-27(in Chinese).

[YJX1] Yin Jing-Xue, On the classical solutions of degenerate quasilinear parabolic equations of the fourth order, J. PDE, 2, No.2 (1985), 39-51.

[YJX2] Yin Jing-Xue, On a class of degenerate quasilinear parabolic equations of fourth order, Northeastern Math. J., 4(1988), 253-270.

[YJX3] Yin Jing-Xue, A uniqueness theorem for a class of degenerate quasilinear parabolic equations of fourth order, J. PDE, 2, No.4, 89-95.

[YJL] Yue Jing-Liang, The Dirichlet problem for a class of quasilinear degenerate elliptic equations, Acta Math. Sinica, 27(1984), 208-222(in Chinese).

[ZK1] Zhang Ke-Nong, The existence of weak solutions of second order differential systems with nonnegative character, Acta Sci. Natr. Univ. Amoien, 21(1982), 14-23(in Chinese).

[ZK2] Zhang Ke-Nong, The uniqueness of weak solutions of second order differential systems with nonnegative characters, Acta Sci. Natr. Univ. Amoien, 22(1983), 110-115(in Chinese).

[ZQ] Zhang Qin, Equilibria of initial-boundary value problem for $u_t = (u^m)_{xx} + (a - x^2)u$, J. PDE, 1A, No.2(1988), 82-96.

[ZJ1] Zhao Jun-Ning, The first boundary-value problem for second order quasilinear elliptic equations with nonnegative characteristic form, Chin. Ann. Math., 4A(1983), 475-686(in Chinese).

[ZJ2] Zhao Jun-Ning, Uniqueness of solutions of the first boundary value problem for quasilinear degenerate parabolic equation, Northeastern Math., 1 (1985), 153-165.

[ZJ3] Zhao Jun-Ning, Applications of the theory of compensated compactness to quasilinear degenerate parabolic equations and quasilinear degenerate elliptic equations, Northeastern Math. J., 2(1986), 33-48.

[ZJ4] Zhao Jun-Ning, Some properties of solutions of quasilinear degenerate parabolic equations and quasilinear degenerate elliptic equations, Northeastern Math. J.,

2(1986), 281-302.

[ZY1] Zhou Yu-Lin, On some problems for nonlinear elliptic equations and nonlinear parabolic equations, J. of Peking Univ. Natur. Sc., 4(1959), 283-326(in Chinese).

[ZY2] Zhou Yu-Lin, Initial value problems for nonlinear degenerate systems of filtration type, Chin. Ann. Math., 5B(1984), 632-652.

[ZQ1] Zhuang Qiong-Shan, The second boundary problem for degenerate parabolic equations, Acta Sci. Natr. Univ. Amoien, 19(1980), 10-24(in Chinese).

[ZQ2] Zhuang Qiong-Shan, The boundary value problem for degenerate elliptic equations, Acta Sci. Natr. Univ. Amoien, 21(1982), 8-13 (in Chinese).

[ZQ3] Zhuang Qiong-Shan, Initial-boundary value problem for double degenerate nonlinear parabolic equation, J. PDE, 2, No. 4 (1989), 47-61.

Nonlinear Hyperbolic Conservation Laws [†]

Xiaxi Ding and Tong Zhang(Tung Chang)

Academia Sinica, Beijing

1. In early 1960's, there were three groups working in this field. They were of Fudan University, Jilin University and Chinese Academy of Sciences. Since the works done in Fudan have been summarized in another paper in this volume and the topics of two spatial dimension case and combustion theory will be talked about in other place, we will give a survey only for the rest in the present paper. We will emphatically talk about the results published in Chinese journals, especially about the isentropic gas flow which has attracted us so much since 1960's.

First, we will introduce nine papers finished in 1963 and their continuiation, most of which are not familiar for westerners. Then some of the works done in 1980's will be mentioned. Our paper will end with the contribution to existence of global big solution for Cauchy problem, the core problem of the whole field, for isentropic flow by using compensated compactness method.

2. It is well known that E. Hopf[1], O.A. Oleinik[2] and P. Lax[3] considered, in 1950's, global discontinuous solutions of Cauchy problem for scalar conservation laws as follows

$$\begin{cases} u_t + f(u)_x = 0, & (x,t) \in I\!\!R \times I\!\!R^+ & (2.1) \\ u(x,0) = u_0(x), & x \in I\!\!R & (2.2) \end{cases}$$

where the flux function is assumed to be convex. i.e.

$$f''(u) \neq 0, \qquad u \in I\!\!R. \tag{2.3}$$

Both existence and uniqueness are solved completely.

In 1960, O.A. Oleinik and A.C. Kalashnikov[4] constructed the local solution for nonconvex case.

In 1963, Z. Wu[5] in Jilin proved the existence of global solution of (2.1-2) for nonconvex case under the assumption that $u_0(x)$ is monotone. He used Douglis[6] framework and successfully extended Douglis ordering principle from convex case to nonconvex.

---

† 1991 Mathematics Subject Classification: Primary 35L65, 35L67; Secondary 65M99

† Partly Supported by National Fundamental Research Program of State Commission of Science and Technology and Academia Sinica

C. Gu et al. (eds.), Partial Differential Equations in China, 19–29.
© *1994 Kluwer Academic Publishers.*

In 1965, L. Lin[7] improved Wu's results to nonmonotone initial data and obtained the first complete existence theorem for Cauchy problem of nonconvex conservation laws. He even proved that the corresponding entropy condition is true at the discontinuity point in solutions. Lin's paper was submitted to the journal Acta Mathematica Sinica. Because of the so-called Cultural Revolution in China, it was not published until 1979. An analogous result was obtained by D. Ballou[8] in 1970.

Suggested by R. Wang, Z. Wu[9] (1963) payed attention to the connection between Finipov's theory of ordinary differential equations with discontinuous right-hand members and the theory of discontinuous solution of conservation laws. He defined the concept of generalized characteristics and then proved the uniqueness of solution of Cauchy problem in the class of piecewise smooth function for nonconvex scalar conservation laws.

3. Let us turn to the system of conservation laws.

In 1860, B. Riemann[10] considered the Cauchy problem with the simplest discontinuous initial data for isentropic flow as follows

$$u_t + p(v)_x = 0, \quad v_t - u_x = 0, \quad (x,t) \in \mathbb{R} \times \mathbb{R}^+ \tag{3.1}$$
$$(u,v)(x,0) = (u^\pm, v^\pm) \quad \text{(arbitrary constants)} \ (x >_> 0) \tag{3.2}$$

where constitution function satisfying $p'(v) < 0$ (hyperbolicity) and $p''(v) > 0$ (convexity). Riemann constructed the solution of (3.1-2). There are four configurations shown in figure 3.1

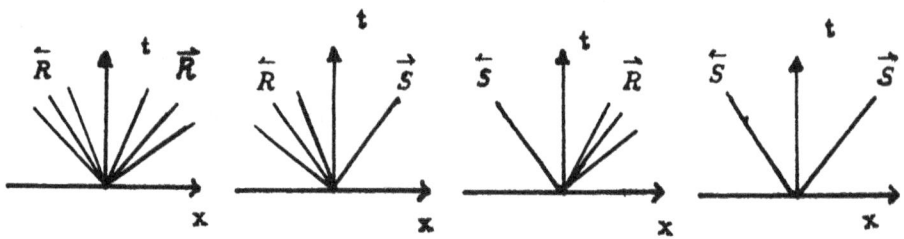

Figure 3.1

where $\overleftrightarrow{R}$ and $\overleftrightarrow{S}$ represent backward or forward rarefaction waves and backward and forward shock waves respectively.

The initial value problem with data consisted of an arbitrary jump and two constant states is called the Riemann problem nowadays[11] and is the most fundamental and active problem in the whole field.

Riemann's result was extended to the following three aspects in China in 1963:

(i) A group of students, directed by S. Ou (Xinmou Wu), extended the initial data possessing two jumps in their bachelor thesis[12]. This problem even has not been solved completely so far, but they did have done some substantial contribution as shown in figure 3.2

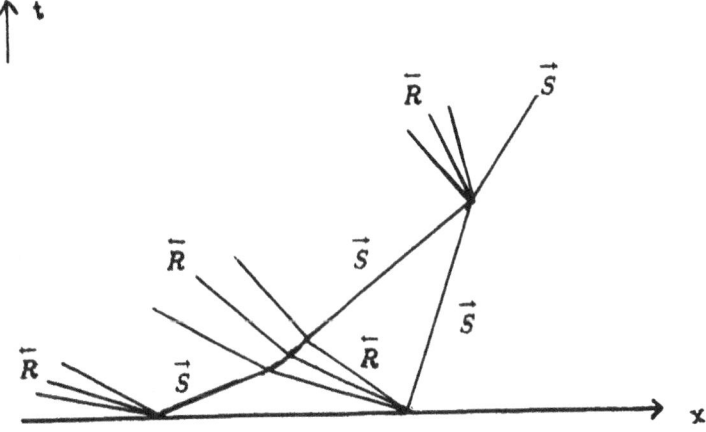

Figure 3.2

Suppose two jumps in initial data project only $\overleftarrow{R}$ and $\overrightarrow{S}$, then the right $\overleftarrow{R}$ and left $\overrightarrow{S}$ collide and penetrate each other. The penetrated $\overrightarrow{S}$ overtakes the right $\overrightarrow{S}$, catchs up to form a transmited $\overrightarrow{S}$ and a reflected $\overleftarrow{R}$. The $\overleftarrow{R}$, will not interact. So, the global solution was obtained. This fact was extended to a class of initial data which only produced forward rarefactive and backward compressive waves by T. Chang and Y. Guo[13] in 1963. Fortunately, [13] was published in 1965 and was announced by S. Ou at an international conference held in Budapest in the same year. Since then, following continuiations have been done:

a. J. Smoller and J. Johnson (1969)[14] extended it to some general 2×2 system under suitable assumption of hyperbolicity and convexity.

b. V. Liapijefskee (1974)[15] proved the uniqueness.

c. X. Ding et. al. (1973)[16] extended the initial data to involve a $\overleftarrow{S}$ additionally. Overtaking of $\overleftarrow{S}$ and $\overleftarrow{R}$ is a special case of the initial data of X. Ding et. al., which is just what C. Moler and J. Smoller (1970)[17] and J. Greenberg (1970)[16] studied independently.

(ii). Directed by X. Ding, J. Wang et. al. (1963) obtained the global smooth solution of isentropic flow with hodogragh method in their bachelor thesis[18]. The earliest global smooth solution was obtained in 1960 by a group directed by C. Gu in Fudan[19] for general 2×2 hyperbolic system

$$r_t + \lambda_1(r, s)r_x = 0, \quad s_t + \lambda_2(r, s)s_x = 0, \tag{3.3}$$

under the assumption that

$$\lambda_1 > \lambda_2 \tag{3.4}$$
$$\lambda_{1r} \geq a > 0, \quad \lambda_{2s} \geq a > 0 \tag{3.5}$$
$$\lambda_{1s} \geq a > 0, \quad \lambda_{2r} \geq a > 0 \tag{3.6}$$

L. Lin (1963) in Jilin clarified that, under the assumption

$$\lambda_{1r} > 0, \quad \lambda_{2s} > 0, \tag{3.7}$$

the necessary and sufficient condition for existence of global smooth solution of (3.3) is

$$r_x(x,0) \geq 0, \quad s_x(x,0) \geq 0. \tag{3.8}$$

The assumption (3.6) is dropped so that (3.3) may involve the isentropic flow model (3.1).

The paper of J. Wang et. al. did not submit to any journal. Lin's result was published in the Acta of Jilin University in Chinese with an English abstract.

Analogous results were obtained independently by P. Lax (1964)[21], M. Yamaguti and T. Nishida (1968)[22].

From then on, Lin consists in studying global smooth solution and obtains many results. The most interesting one of them is his special contribution to vacuum problem. Lin (1987)[23] showed that vacuum states can not appear in rarefaction solutions of (3.1) unless the vacuum is present as time $t = 0+$.

iii. I. Gelfand (1959)[24] constructed the solution of Riemann problem for non-convex scalar conservation law (2.1) with the entropy condition at jumps:

$$\frac{f(u) - f(u_l)}{u - u_l} \geq \frac{f(u_r) - f(u_l)}{u_r - u_l} \quad \forall u \in \{u \mid (u^- - u)(u^+ - u) \leq 0\} \tag{3.9}$$

where u_l and u_r denote the limits of u at left and right banks at jumps in the solution respectively.

T. Zhang found that the Riemann solution can be expressed as

$$u\left(\frac{x}{t}\right) = \begin{cases} u^{\pm}, & \frac{x}{t} \bar{\in} (\tilde{f}'(u^-; u^{\pm}), \tilde{f}'(u^+; u^{\pm})) \\ \tilde{f}^{-1}(u, u^{\pm}), & \frac{x}{t} \in (\tilde{f}'(u^-; u^{\pm}), \tilde{f}'(u^+; u^{\pm})) \end{cases} \tag{3.10}$$

where $\tilde{f}(u; u^{\pm})$ is the convex hull of $f(u)$ through points $(u^{\pm}, f(u^{\pm}))$ in phase plane (u, f).

Directed by T. Zhang with his suggestion to remove (3.9-10) to backward and forward waves of (3.1) separately, C. Li et. al. (1963) extended Riemann's result to nonconvex case in their Bachelor thesis [25]. Furthermore, T.Zhang and L. Hsiao (1963)[26] extended [25] to general Riemann problem

$$(u, v)(x, 0) = (u_0^{\pm}(x), v_0^{\pm}(x)) \quad (x \overset{>}{<} 0) \tag{3.11}$$

and constructed the local solutions. [26] is also an extension of [27] finished in Fudan, which solved (3.1) and (3.11) for convex case. The paper of Zhang and Hsiao was submitted to Acta Math. Sinica in 1963, and an abstract of four pages was published in 1977. The English translation of the whole paper with a little revisements was published in 1981[29].

Analogous result to [25] was obtained by B. Wendroff (1972)[29] and L. Leboviech (1974)[30] independently. The entropy condition for general nonconvex system was suggested by T.P. Liu (1974)[30] L. Hsiao and T.Zhang (1978)[31] extended [25] to a kind of general nonconvex 2×2 system under suitable hyperbolicity assumption independently.

Zhang, Li and Hsiao (1975)[32] studied interaction of waves for nonconvex case, and successfully droped the convex assumption in Zhang and Guo's result.

Besides, Z. Wu (1963)[33] proved the uniqueness of Cauchy problem for isentropic flow in the class of piecewise smooth functions with centered rarefaction waves. It is an improvement of O.A. Oleinik's result [34].

Combining the iteration procedure used by Courant and Lax for Cauchy problem and the technique of Hartman and Wintner for estimating continuous modulus, R. Wang and Z. Wu (1963)[35] establish the classical theory for smooth local solution of various initial boundary value problems for $n \times n$ quasilinear system with two independent variables.

These are the results about the nine papers finished in 1963 in China.

4. The Riemann problem for one-dimensional adiabatic flow is described as follows

$$\begin{cases} u_t + p_x = 0, \\ v_t - u_x = 0, \\ E_t + (pu)_x = 0, \ E = e + \frac{u^2}{2}, \ e = \frac{pv}{\gamma-1}, \ \gamma > 1 \end{cases} \tag{4.1}$$

$$(u,p,v)(x,0) = (u^\pm, p^\pm, v^\pm) \ (x \overset{>}{<} 0) \tag{4.2}$$

Besides $\overrightarrow{R}, \overrightarrow{S}$, the elementary wave of (4.1) involves contact discontinuity $\overset{<}{\overset{>}{J}}$: $x = 0$, $u_l = u_r$, $p_l = p_r$, $v_l \overset{>}{<} v_r$. (4.1-2) was solved by Courant and Friedrichs[36]. Moreover, they obtained the fundamental results of interaction of elementary waves, to which some supplement was contributed by L. Hsiao and T. Zhang[37] in 1979.

An extension for nonconvex case of (4.1-2) was obtained by Zhang and Hsiao[38] in 1979 in the case of $e = pf(v)$, $f(v) > 0$, $f'(v) \geq 0$, $\lim_{v \to 0} f(v) = 0$ and $f''(v) \leq 0$ as v being sufficiently large. Furthermore, [38] clarified that the solution possesses the maximum of entropy in a certain sense.

B. Wendroff[29] (1972) and T.-P. Liu[39] (1975) studied (4.1-2) in nonconvex case also.

Most of the Chinese materials mentioned in section 3-4 are collected in the monogragh "The Riemann Problem and Interaction of Waves in Gas Dynamics" written by Chang(Zhang) and Hsiao(Xiao)[40].

5. Since 1979, the situation in China has been changing a lot. Some of Chinese mathematicians have chance to visit western countries and to do cooperative work with westerners. Many new results have been obtained. We will mentioned some of them as follows.

Entropy theory is one of the most interesting problems in this field. For guaranteeing uniqueness, the discontinuous solution has to satisfy some entropy criterion. There are several entropy criteria and one of them is so-call entropy rate admissibility criterion suggested by C. Dafermos[42] (1979). L. Hsiao[43] (1980) proved that this criterion is equivalent to others when $\gamma \geq 5/3$ (monatomic gas) but not equivalent when $\gamma < 5/3$ and the shocks are very strong. This is a very interesting contribution to entropy theory.

L. Ying and J. Wang[44] (1980) proved the existence of global solutions of Cauchy problem for system (2.1) with inhomogeneous term on which only rather simple restriction is assumed. A generalization of Glimm's method[45] is proposed in the proof. This is one of the earliest result for inhomogeneous system of conservation laws, and the other earliest works are due to T.P. Liu[46] (1979) and C. Dafermos and L.Hsiao[47] (1982).

J. Wang and C. Li[48] (1982) proved the existence of global solutions of general Riemann problem (3.11) for isentropic flow of polytropic gas provided that

$$\frac{\gamma - 1}{2}(|u_0(0+) - u_0(0-) + |v_0(0+) - v_0(0-)|)(\underset{x<0}{\mathrm{TV}}\{u_0, v_0\} + \underset{x>0}{\mathrm{TV}}\{u_0, v_0\})$$

is sufficiently small. It is also showed that the structure of the solution is similar to the corresponding Riemann solution

$$(u, v)(x, 0) = (u_0(0\mp), v_0(0\mp))\ (x \overset{<}{>} 0)$$

In other words, it means that the Riemann solution is globally stable under small perturbation of initial data.

T.-P. Liu and J.Wang[49] (1985) proved the existence of global solutions for

$$\begin{cases} u_t + (\phi u)_x = 0, \quad v_t + (\phi v)_x = 0 & (5.1) \\ (u, v)(x, 0) = (u_0(x), v_0(x)) & (5.2) \end{cases}$$

where $\phi = \phi(u, v)$. System (5.1) is derived as a model for elastic string[50] and is nonstrictly hyperbolic.

More interest is that they clarified behavior of solutions in large time, and nonstrict hyperbolicity may cause stronger nonlinear interactions between waves pertaining to different families.

C. Li and T.-P. Liu[51] (1983) studied the asymptotic states for hyperbolic conservation laws with a moving source. They construct noninteracting wave patterns (i.e. asymptotic states). When nonlinear resonance occurs, instablility may result. They identify a stability criterion which is independent of the flux function (even nonconvex). The scalar model as well as transonic gas flows through a duct with varying cross section is studied. For the latter case, it is shown that the stability of a wave pattern depends on the geometry of the duct, and not on the equation of the state. In particular, transonic steady shock waves along a converging duct are unstable, and flow along a diverging duct is always stable.

6. Compensated compactness theory and hyperbolic conservation laws

It is well known that the Sobolev space is successfully used in various topics of partial differetial equations except in the theory of discontinuous solution of hyperbolic conservation laws, because the discontinuous solutions do not have generalized derivatives in the sense of Sobolev. But the circumstance has been changed since the appearance of compensated compactness theory in early 1980's[52-3]. In this theory, people use Sobolev space of negative order $H^{-1}(\Omega)$. Using compensated compactness theory X. Ding, G. Chen and P. Luo[54-6] (1987-1989) proved the convergence of the Lax-Friedrichs and Godunov difference schemes and then obtained the global existence theorem of Cauchy problem for equations of isentropic flow of polytropic gas with adiabatic exponent $\gamma \in (1, 5/3]$. The special case of the theorem as $\gamma = 1 + \frac{2}{2m-1}$, $m(\geq 2)$ being integers was obtained before by R. Diperna[53] in 1983 with viscosity method. The main tool used in [54-6] is the following

Compensated compactness theorem:

Let $\Omega \subset \mathbb{R}^+ \times \mathbb{R}$ be a bounded open set, $u^\epsilon : \Omega \longrightarrow \mathbb{R}^4$, $u^\epsilon = (u_1^\epsilon, u_2^\epsilon, u_3^\epsilon, u_4^\epsilon)$ are measurable,

$u^\epsilon \rightharpoonup u$ in $L_4^2(\Omega)$, ("$\rightharpoonup$" denotes weak convergence)

$$\begin{cases} \dfrac{\partial u_1^\epsilon}{\partial t} + \dfrac{\partial u_2^\epsilon}{\partial x} \\ \dfrac{\partial u_3^\epsilon}{\partial t} + \dfrac{\partial u_4^\epsilon}{\partial x} \end{cases} \quad \text{are compact in} H_{loc}^{-1}(\Omega),$$

then we can get a subsequence (denoting it by u^ϵ again) such that

$$\begin{vmatrix} u_1^\epsilon & u_2^\epsilon \\ u_3^\epsilon & u_4^\epsilon \end{vmatrix} \longrightarrow \begin{vmatrix} u_1 & u_2 \\ u_3 & u_4 \end{vmatrix} \quad \text{in the sense of distribution.}$$

In the proof, we used the fact that

$$\frac{1}{2} \begin{vmatrix} a & b \\ c & d \end{vmatrix}$$

denotes the area of the triangle with vertices (a, b), (c, d), $(0, 0)$ and is invariant under any rotation around the origin.

For general 2×2 system of conservation laws

$$u_t + f(u)_x = 0,$$

we call $(\eta(u), q(u))$ the Lax entropy pair if they satisfy

$$\eta(u)_t + q(u)_x = 0.$$

So we have

$$\nabla q = \nabla \eta \nabla f.$$

Eliminating q, we can get a linear second order equation of $\eta(u)$. In the case of isentropic gas dynamic system

$$\begin{cases} \rho_t + m_x = 0, \quad m = \rho u \\ m_t + (\frac{m^2}{\rho} + p(\rho))_x = 0, \quad p = e^r, \ r \in (1, 5/3) \end{cases}$$

the second order equation satisfied by $\eta(u)$ is the Euler-Poisson equation

$$\eta_{wz} - \frac{\lambda}{w-z}(\eta_w - \eta_z) = 0, \qquad \lambda = \frac{3-\gamma}{2(\gamma-1)},$$

where w, z are the so-called Riemann invariants.

For general weak entropy $\eta(w, z)$ we have the representation formula

$$\eta(w, z) = \int_z^w (w - s)^\lambda (s - z)^\lambda \phi(s) ds,$$

where $\phi(s)$ is arbitrary. Suitably choosing $\phi(s)$, we can get a sequence $\eta_n(w, z)$ of weak entropies. In addition, using the Young representation theorem of Radon measure and very complicated estimations related to $\eta_n(w, z)$ finally we proved the existence of global solution of Cauchy problem above.

G. Chen and Y. Lu[57] (1988) proved the existence theorem of a scalar conservation law by compensated compactness method without using measure theory.

References

[1] Hopf, E. (1950). The partial differential equation $u_t + uu_x = \mu u_{xx}$. Comm. Pure Appl. Math. 3:201-230.

[2] Olienik, O.A. (1957). Discontinuous solution of nonlinear differential equations. Usp. Mat. Nauk 12:3-73. English Trans. in Amer. Soc. Transl. Ser. 2, 26:95-172.

[3] Lax, P. (1954), Weak solutions of nonlinear hyperbolic equations and their numerical computation. Comm. Pure Appl. Math. 7:159-193.

[4] Oleinik, O.A. & Kalashnikov, A.C. (1960) A class of discontinuous solutions for first order quasilinear equations. Proc. Conf. Diff. Equat. 133-137, Jerevon (in Russian).

[5] Wu, Zhuoqun (1963). Existence and uniqueness of generalized solution of Cauchy problem for first order quasilinear differential equation without convexity condition, Acta Math. Sinica, 13:515-530 (in Chinese).

[6] Douglis, A., (1959), An ordering principle and generalized solutions of certain quasi-linear partial differential equations, Comm. Pure Appl. Math., 12:87-112.

[7] Lin, Longwei (1979), The global solutions of Cauchy problem for quasi-linear hyperbolic equation without convexity condition, Acta Jilin Univ., 2:17-26. (in Chinese with English abstract)

[8] Ballou, D. (1970), Solution to nonlinear hyperbolic Cauchy problems without convexity conditions, Tran. AMS. 152:441-460.

[9] Wu, Zhouqun (1963), The ordinary differential equations with discontinuous right-hand members and the discontinuous solutions of the quasilinear partial differential equations. Acta Math. Sinica 13:515-30. English trans. in Sciential Sinica 13:1901-1907.

[10] Riemann, B. (1896), Desammeltte Werke.

[11] Lax, P. (1957), Hyperbolic system of conservation laws II. Comm. Pure Appl. Math. 10:537-566.

[12] Zhang, Chenjun; Sun, Geng; Xiong, Jiwu; Liao, Yingxin; Jia, Shufeng and Yang, Tongzu (1963). Initial value problem with three constant states as initial data for

an aerodynamic system. Bachelor thesis of the Chinese Science and Technology Univ.

[13] Chang, Tong and Guo, Yufa (1965). A class of initial value problems for system of aerodynamic equations. Acta Math. Sinica 15:386-96; English trans. in Chinese Math. 7(1965):90-101.

[14] Smoller, J. and Johnson, J. (1969). Global solutions for an extended class of hyperbolic system of conservation laws. Arch. Rat. Mech. Anal. 32:169-189.

[15] Liapijefskee, V. (1974). On uniqueness of generalised solution for aerodynamic equations. Dokl. Akad. Nauk. SSSR 215:535-538 (in Russian).

[16] Ding, Xiaxi; Chang, Tung; Wang, Chinghua; Hsiao, Ling and Li, Caizhong (1973), A study of the global solutions for a quasilinear hyperbolic system of conservation laws, Scientia Sinica 16:317-335.

[17] Moler, C. and Smoller, J. (1970). Elementary interaction in quasilinear hyperbolic systems. Arch. Rat. Mech. Anal. 37:309-322.

[18] Greenberg, J. (1970). On the interaction of shocks and simple waves of the same family. Arch. Rat. Mech. Anal. 37:136-160.

[19] Gu. Chaohao et. al. (1962), A global solution without shocks for quasilinear hyperbolic system of equations. Math. Anthology of Fudan Univ. 36-39 (in Chinese).

[20] Lin, Longwei (1963). On the global existence of the continuous solutions of the reducible quasilinear hyperbloic system. Acta Jilin Univ. 4:83-96.

[21] Lax, D. (1964). Development of singularities of solutions of linear hyperbolic partial differential equations. J. Math. Phys. 5:611-613.

[22] Yamaguti, M. and Nishida, T. (1968). On some global solutions for quasilinear hyperbolic equations. Funkcialaj Ekvacioj 11:51-57.

[23] Lin, Longwei (1987). On the vacuum state for the equations of isentropic gas dynamics, J. Math. Anal. Appl. 121:406-425.

[24] Gelfand, I. (1959). Some problems in the theory of quasilinear equations, Usp. Mat. Nauk. 14:87-158 (in Russian).

[25] Li, Caizhong; Hsiao, Ling; Yang, Shaoqi and Yuan, Zuwen (1963), Riemann problem for typical quasilinear hyperbolic system without convexity. Bachelor thesis of the Chinese Science and Technology Univ.

[26] Chang, Tung and Hiao, Ling (1977). Riemann problem and discontinuous initial-value problem for typical quasilinear hyperbolic system without convexity. Acta Math. Sinica 20:229-231 (in Chinese).

[27] Gu. Chaohao; Li, Daqian; Yu, Wenci and Hou, Zongyi (1961-1962), Discontinuous initial-value problem for hyperbolic system of quasilinear equations (I), (II), (III). Acta Math. Sinica 11:314-323; 324-337; 12:132-143 (in Chinese).

[28] Hsiao, Ling and Zhang, Tong (1981). Perturbation of the Riemann problem in gas dynamics. J. Math Anal. Appl. 79:436-460.

[29] Wendroff, B. (1972). The Riemann problem for materials with nonconvex equations of state I. Isentropic Flow; II. General flow. J. Math. Anal. Appl., 38:454-466: 640-658.

[30] Liu, T.-P. (1974), The Riemann problem for general 2×2 conservation laws. Trans. AMS. 199:89-112.

[31] Hsiao, Ling and Zhang, Tong (1978). Riemann problem for 2×2 quasilinear hyperbolic system without convexity. Kexue Tongbao 8:465-469 (in Chinese).

[32] Chang, Tung; Li, Caizhong and Hsiao, Ling (1975) Global solution for a class of initial-value problem of a typical quasilinear hyperbolic system without convexity. Kexue Tongbao, 20:506-510 (in Chinese).

[33] Wu, Zhouqun (1963). Uniqueness for discontinuous solution with centered rarefaction waves for aerodynamic equations. Acta Jilin Univ., 4:35-49 (in Chinese with an English abstract).

[34] Oleinik, O.A.(1957). Uniqueness of generalized solutions of Cauchy problem for a quasilinear system of equations appearing in mechanics. Usp. Mat. Nauk. 11:169-176. (in Russian).

[35] Wang, Rouhwai and Wu, Zhuoqun (1963). On mixed initial boundary value problem for quasi-linear hyperbolic equations in two independent variables. Acta Jilin Univ. 2:459-502 (in Chinese with an English absract).

[36] Courant, R. and Friedrichs, K.O. (1948). Supersonic Flow and Shock Waves. Wiley-Interscience, New York.

[37] Hsiao, Ling and Zhang, Tong (1979). Interaction of elementary waves in one-dimensional adiabatic flow. Acta Math. Sinica 22:596-619 (in Chinese with an English abstract).

[38] Chang, Tung and Hsiao, Ling (1979). Riemann problem for one-dimensional adiabatic flow without convexity. Acta Math. Sinica 22:229-231 (in Chinese with an English abstract).

[39] Liu, T.P. (1975). Existence and uniqueness theorems for Riemann problem. Trans. AMS. 213:3755-3782.

[40] Chang, Tong and Hiao, Ling (1989). The Riemann problem and interaction of waves in gas dynamics, Pitman Monographs and Surveys in Pure and Appl. Math. 41, Longman Scien. Tech.

[42] Dafermos, C. (1979). The entropy rate admissibility criterion for solutions of hyperbolic conservation laws. J. Diff. Equat. 14:202-212.

[43] Hsiao, Ling (1980). The entropy rate admissibility criterion in gas dynamics, J. Diff. Equat. 38:226-238.

[44] Ying, Lungan and Wang, Chinghua (1980). Global solutions of the Cauchy problem for a nonhomogeneous quasilinear hyperbolic system. Comm. Pure Appl. Math. 33:579-597.

[45] Glimm, J. (1965). Solutions in the large for nonlinear hyperbolic systems of equations. Comm. Pure Appl. Math. 18:697-715.

[46] Liu, T.-P. (1979), Quasilinear hyperbolic system, Comm. Math. Phys., 68:141-172.

[47] Dafermos, C. and Hsiao, Ling (1982). Hyperbolic system of balance laws with inhomogeneity and dissipation. Indiana Univ. Math. J. 331:471-491.

[48] Wang, Chinghua and Li, Caizhong (1982). Study of global solutions for nonlinear conservation laws. J. Math. Anal. Appl. 85:236-256.

[49] Liu, T.-P. and Wang, Chinghua (1985), On a nonstrictly hyperbolic system of conservation laws, J.Diff. Equat., 57:1-14.

[50] Keyfitz, B. and Kranzer, H. (1980), A system of nonstrictly hyperbolic conservation laws.

[51] Li, Caizhong and Liu, T.-P. (1983), Asymptotic states for hyperbolic conservation laws with a moving source. Advances Appl. Math., 4:353-379.

[52] Tartar, L. (1979) Compensated compactness and applications to partial differential equations. Research Notes in Math., nonlinear analysis and mechanics: Heriot Watt Symp., Vol. 4 Knods, R.J. (ed.), New York, Pitman Press.

[53] Diperna, R. (1983). Convergence of the viscocity method for isentropic gas dynamics. Comm. Math. Phys., 91:1-30.

[54] Ding, Xiaxi; Chen, Guiqiang and Luo, Peizhu (1987-8) Convergence of the Lax-Friedrichs scheme for isentropic flow, Acta Math. Sinica (I) 7:483-540; (II) 8:61-94. (in Chinese).

[55] Chen, Guiqiang (1988), ibid (III) 8:101-134 (in Chinese).

[56] Ding, Xiaqi; Chen Guiqiang and Luo, Peizhu (1989) Convergence of the fractional step Lax-Friedrichs and Godunov scheme for the isentropic system of gas dynamics. Comm. Math. Phys., 121:63-84.

[57] Chen, Guiqiang and Lu, Yungguang (1988). Study of application route of compensated compactness theory, Kexue Tongbao, 33:641-644 (in Chinese).

Elliptic and Parabolic Equations[†]

Guangcang Dong[††]

Department of Mathematics, Zhejiang University, Hangzhou

: **Abstract:** This article provides a brief servey of the results on the partial differential equations and systems of elliptic and parabolic types contributed by the Chinese mathematicians in the period from 1950 to 1980. The major contributions in this period are summarized in the following topics.

1. On the definition of elliptic systems.
2. New prototypes of initial and boundary value problems
3. Schauder estimates for solutions to high order parabolic equations and systems
4. Boundary value problems of first order elliptic systems with two variables via the theory of generalized analytic functions
5. Linear boundary value problems of the other types
6. Asymptotic behavior of solutions to parabolic equations
7. Nonlinear parabolic equations of second order
8. Calcilus of variations with strong nonlinearities
9. Subsonic flows in high dimensions

In the field of elliptic and parabolic equations (or systems), the early researches of Chinese mathematicians laid emphasis on classification of equations and various linear and nonlinear prescribed boundary value problems. A brief introduction is as follows.

1   On the definition of ellipticity for systems

Suppose A, B, C are real square matrices of 2-order. Consider the partial differential system:

$$\mathcal{L}(u) \equiv \left(A\frac{\partial^2}{\partial x^2} + 2B\frac{\partial^2}{\partial x \partial y} + C\frac{\partial^2}{\partial y^2}\right)\binom{u}{v} = 0. \tag{1}$$

By the definition of I.G. Petrovskii, if

$$|A + 2B\lambda + C\lambda^2| \neq 0, \quad \forall \lambda \in \mathbb{R},$$

† 1991 Mathematics Subject Classification: 35J25, 35J65, 35K60
†† Supported by NNSF of China and Foundation of Science of Zhejiang Province

C. Gu et al. (eds.), Partial Differential Equations in China, 30–41.
© 1994 Kluwer Academic Publishers.

then (1) is called an elliptic system. One of the benefits of this definition is that normal solution of elliptic systems are analytic. But there are examples showing that this definition can't assure the uniqueness of solutions for Dirichlet problems. In [1], I.M. Visik introduced the strongly ellipticity condition, that is

$$\tilde{A} + 2\tilde{B}\lambda + \tilde{C}\lambda^2 > 0, \quad \forall\lambda \in \mathbf{R},$$

where $\tilde{D}$ denotes the symmetric part of an square matric D. He also proved that this condition is sufficient for uniqueness of Dirichlet problems. But some examples show that this condition is not necessary.

In 1960, Ding X.X. et al proved, in [2], that the sufficient and necessary condition for uniqueness of the Dirichlet problems is the following

$$|A + 2Bb + Cc| \neq 0, \quad \forall b^2 \leq c. \tag{D}$$

The necessity is easy to know. Their method of proving sufficiency is as follows.

1. In general, there exists a nonsingular square matrix P such that PL(u) is a strongly elliptic system.

2. In some special cases, one can find a singular matrice P such that the energy integral method works for PL(u) to yield the uniqueness.

Later L. G. Hua et al proved, in [3], that the condition (D) is equivalent to the following:

There exist nonsingular square matrices P and Q such that
PL(u)Q=0 is a strongly elliptic system.

They reduced PL(u)=0 into one of four standard forms and proved the uniqueness of Dirichlet problems by using energy integral method. In the monograph [4] and their following researches, Hua et al gave a complete classification for system (1) and also obtained series of results for boundary value problems.

2 New prototypes of initial and boundary value problems

For harmonic equation $\Delta u = \sum_{i=1}^n \frac{\partial^2 u}{\partial x^2} = 0 (x \in \Omega)$ or generalized elliptic equations, we usually have Dirichlet condition $u|_{\partial\Omega} = \phi(x)(\phi(x)(x \in \partial\Omega)$ is a given function) or Neuman condition $\sigma\frac{\partial u}{\partial N}|_{\partial\Omega} = \phi$ (σ, ϕ are given functions on $\partial\Omega$ and N is the outer normal of $\partial\Omega$) or oblique derivative condition $\sigma\frac{\partial u}{\partial\nu} + \lambda u = \phi$ (σ, λ, ϕ are given functions on $\partial\Omega$, ν is a given oblique direction variant with the points on $\partial\Omega$. We usually have $cos(\nu, N) \neq 0$).

Li D.Q. et al are the first who put forth the boundary value conditions on equal value surface. Assume Ω is a multiply connected region, $\partial\Omega = \partial\Omega_0 \cup \sum_{i=1}^n \partial\Omega_i$, where $\partial \Omega_0$ is the outer boundary, $\partial \Omega_i \cap \partial \Omega_j = \emptyset$ ($\emptyset$ is the empty Ω set, $0 \leq i, j \leq n$). One has boundary conditions on $\partial \Omega (i \geq 1)$: $u = c_i$, $\int_{\partial \Omega} \sigma \frac{\partial u}{\partial N} ds = A_i$. Here c_i are constants to be determined, A_i are fixed constants. On $\partial \Omega_0$ one still has Dirichlet condition. The existence and uniqueness of the solutions and their dependence continuously on the boundary data were proved. These boundary conditions arise in physics in [5] and [6] by using Green formula and other methods. In [7] the above results were extended to complementary boundary conditions. Furthermore, in [8], Chen S.X. extended the correspoding initial and boundary value conditions and obtained some interesting results.

3 Schauder estimates for parabolic equations and systems of high order

Schauder estimates of solutions, which include interior estimates and estimates near the boundary, play an important role in studying eliptic and parabolic equations (or systems). The estimates for elliptic equations and systems have been summarized in [9], in which the estimates for elliptic equations and systems of high order satisfying general boundary conditions were obtained. The interior estimates of parabolic equations and systems are contained in Friedman [10], in which there is no essential difference between second order and high order equations. The estimates near the boundary for second order equations were established in Friedman [11]. Complete estimates for general parabolic equations and systems with general boundary conditions were obtained in Teng Z.H. [12] and Wang R.H. [13].

Now we state the assumptions and results obtained in [12]. (For simplicity, we treat here a single equation.)

Let $P = (x,t)$, $Q = (\xi, \tau)$ be two points in $\mathbb{R}^{n+1} \times \mathbb{R}^+$. The distance between them is defined by $|p - Q| = (|x - \xi|^2 - |t - \tau|^{\frac{1}{m}})^{\frac{1}{2}}$, where m is a fixed positive integer.

Let $K = (k_0, ..., k_{n+1})$, where $k_j (0 \leq j \leq n+1)$ are nonnegative integers, $k = |K| = 2mk_0 + \sum_{j=1}^{m+1} k_j$. We denote $\tilde{D}^k = \frac{\partial^{k_0}}{\partial t^{k_0}} D_{x_1}^{k_1} ... D_{x_{n+1}}^{k_{n+1}}$, $D_{x_j} = \frac{1}{i} \frac{\partial}{\partial x_j} (1 \leq j \leq n+1)$, where $i = \sqrt{-1}$.

Given a function f=f(x,t) defined in some domain $A \subset \mathbb{R}^{n+1} \times \mathbb{R}^+$, we define the following seminorms:

$$[f]_k^A = max\{ \sup_{P \in A, |K| = \|} |\tilde{D}^k f(P)|, \sup_{\substack{p = (x,t) \in A, \\ \Omega = (\xi, \tau) \in A \\ max\{k-2m,0\} \\ \leq |s| = s < k}} \frac{|\tilde{D}^k f(P) - \tilde{D}^s f(Q)|}{|P - Q|^{k-s}} \}$$

$$[f]_{k+\alpha}^A = max\{ \sup_{P \in A, |K| = \|} |\tilde{D}^k f(P)|, \sup_{\substack{p = (x,t) \in A, \\ \Omega = (\xi, \tau) \in A \\ max\{k-2m,0\} \\ \leq |s| = s < k}} \frac{|\tilde{D}^k f(P) - \tilde{D}^s f(Q)|}{|P - Q|^{k-s}} \}$$

where $0 < \alpha < 1$.

We define

$$|f|_k^A = \sum_{j=0}^{k} [f]_j^A, \qquad |f|_{k+\alpha}^A = |f|_k^A + [f]_{k+\alpha}^A$$

Replacing x_{n+1} by y we denote $p = (x_1, ..., x_n, y)$, $D = (D_{x_1}, ..., D_{x_n}, D_y)$.

Let's consider the following equation with variable coefficients.

$$M(P, \tilde{D})u(P) = F(P), \quad in \ D \subset \{(\S, \dagger, \sqcup)| \imath < \sqcup < \Upsilon, \ \dagger > \imath\},$$
$$B_j(x, t, \tilde{D})u(P) = \Phi_j(x, t), \quad on \ the \ surface \ \Gamma \ of \ D,$$
$$u(P) = \Phi(x, y), \quad on \ the \ base \ \omega \ of \ D,$$

where M is a differential operator of order 2m and B_j of order m_j. The coefficients of B_j do not depend on y. M, B_j may be written in more explicit form:

$$M = \frac{\partial}{\partial t} - L(P, D_x, D_y),$$
$$L = \sum_{|k| \leq 2m} a_k(P) D^k$$
$$B_j = \sum_{|\gamma| \leq m_j} b_{j,\gamma}(x, t) \tilde{D}^\gamma$$

The leading parts of M, B_j (the parts of highest order) are taken to be M', B_j' respectively.

Let $M' = \frac{\partial}{\partial t} - L'$. We assume that the following conditions are satisfied:

(i) The parabolic condition of M: The coefficients of M' are real and M' is uniformly parabolic, i.e., there is a constant $A > 0$ such that every root p of the polynomial $M'(P, p, \alpha, \beta)$ satisfies the inequality

$$-A(|\alpha|^2 + |\beta|^2)^m \leq Rep \leq -A^{-1}(|\alpha|^2 + |\beta|^2)^m$$

for any real vector $(\alpha, \beta) = (\alpha_1, ..., \alpha_n, \beta)$ and $P \in D$.

(ii) For any fixed point $P^* = (x^*, t^*, 0)$, the system with constant coefficients, which consists of operators $M'(P^*, \tilde{D})$ and $B_j'(P^*, \tilde{D})(1 \leq j \leq m)$, satisfies a uniformly complete condition with respect to P^*. Note that $M'(P^*, p, \alpha, \beta)$ is a 2m-order polynomial in β with real coefficients, which has m roots β_k^+ with positive imaginary parts and m roots β_k^- with negative imaginary parts $(1 \leq k \leq m)$. It follows that

$$M'(P^*, p, \alpha, \beta) = \prod_{k=1}^{m} \{[\beta - \beta_k^+(P^*, p, \alpha)][\beta - \beta_k^-(P^*, p, \alpha)]\}$$

We denote

$$M^*(P^*, p, \alpha, \beta) = \prod_{k=1}^{m} [\beta - \beta_k^+(P^*, p, \alpha)].$$

Let $B_j(P^*, p, \alpha, \beta) = \sum_{k=1}^{m} b_{j,k}(P^*, p, \alpha, \beta)\beta^{k-1} (mod M^+)$.

The complete condition means $d(P^*, p, \alpha) = det(b_{j,k}(P^*, p, \alpha)) \neq 0$ for $(p, \alpha) \in W' = \{(p, \alpha)|Rep > 0, |p| + |\alpha| \neq 0\}$. The uniformly complete condition means that $\min_{(p,\alpha)\in W', |p|^{\frac{1}{m}} + |\alpha|^2 = 1} d(P^*, p, \alpha)$ has a positive lower bound independent of P^*.

Then under some smoothness conditions on the boundary Γ, the coefficients of the equation, and the boundary data, the author proved

$$|u|_{l+\alpha}^{D} \leq C(|F|_{l-2m}^{D} + \sum [\Phi_j]_{l-m_j+\alpha}^{\Gamma} + [\Phi]_{l+\alpha}^{\omega} + |u|_0^{D})$$

for fixed $l \geq \{m, m_1, ..., m_n\}$.

The above inequalities are Schauder estimates in the entire domain. The estimates in a small portion near the boundary are not hard to derive.

The proof of the above results was carried out in the standard way. A Green function and its precise estimates of half-space problem were given for the equation with constant coefficients. With the explicit representations of solutions in terms of the fundamental solution and Green functions and also potiential theory, the estimates of the solutions were obtained. Finally, these estimates were generalized to equations with variable coefficients.

4   Two variables elliptic systems of first order by the method of generalized analytic function

The theory of generalized analytic function is crucial method in study of the two variables (linear or nonlinear) elliptic systems of first order. The paper in this aspect, we refer to [14]-[18] etc. Here we only illustrate one result related to this topic. Li Z. and Wen G.C. [15], [16] had studied the general type of two variables elliptic systems:

$$-v_y + a_{11}u_x + a_{12}u_y + b_{12}v = f,$$
$$v_x + a_{21}u_x + a_{22}v = g, \tag{2}$$

where $a_{11} > 0$, $a_{11}a_{22} - \frac{1}{4}(a_{12} = a_{21})^2 \geq \Delta_0 > 0$. By transformation of the variables, it yields the complex form:

$$\omega_{\bar{z}} - Q(z)\overline{\omega}_{\bar{z}} = A(z)\omega + B(z)\overline{\Omega} + C(z) \tag{3}$$

Assume $Q(z)$ is bounded and measurable in a bounded domain D and Hölder continuous in a neighborhood of ∂D, and also satisfies the uniform elliptic condition in $D + D_{\partial D}$, where $D_{\partial D}$ is a small neighborhood of ∂D. $|Q(z)| \leq Q_0 < 1$, for some constant Q_0. Moreover we assume A, B, C $\in L_p(D)$, $p > 2$. We want to find solutions of (3) in D of the type $\omega(z) = u + iv$ such that $\omega(z)$ is continuous on $D \cup \partial D$ and satisfies the boundary condition

$$Re[\overline{\lambda}(s), \omega(s)] = \gamma(s), \quad s \in \partial D, \tag{4}$$

where $\lambda(s)$, $\gamma(s) \in C_r(\partial D)(0 < r \leq 1)$ are prescribed functions. It is the well-known Riemann-Hilbert problem. In [15][16], they had given the existence condition, the number and the representative formula of solutions. Their proof contributes to the study of singular integral equations.

5   Other results of linear equations

1. The topic about extension of the maximum principle to elliptic equations of high order of special type we refer to Lu W.R. [10] etc. His result is also a extension of some results of foreign scholars.

Let $\Omega \subset \mathbb{R}^n$ be a bounded domain, $P(x)$, $Q(x)$ satisfy $fg > 0$, $(fg)' > 0$. The equation

$$\Delta(Q(x)g(x)\Delta u) + P(x)f(u) = 0$$

satisfying $\Delta u|_{\partial\Omega} = 0$, then there exists $x_0 \in \partial\Omega$ such that $u(x) \leq u_0(x)$ for all $x \in \Omega$.

Above theorem and other analoguous results can be used to prove the existence and uniqueness of some equations with prescribed boundary value. The proof of maximum principle uses the method of comparison of function.

2. In Lu L.J. [20], he proved uniqueness of Dirichlet problem of general linear elliptic equations with prescribed measure on the boundary $\partial\Omega$, where $\Omega \subset \mathbb{R}^n$ is a bounded domain, and general linear parabolic equations in $\Omega \times \mathbb{R}^n$ with prescribed initial surface and boundary surface measure. His proof used the method of barrier function.

3. In Shen Y.T. [21], he studied the existence and uniqueness of degenerate or singular elliptic equations of high order. His proof used the embedding theorem of weighted function.

4. The asymptotic property of parabolic equations.

In Gu L.K. [22][23], he studied the asymptotic property of Cauchy problem of linear and semi-linear parabolic equations. Consider

$$\mathcal{L}u = \sum_{i,j=1}^{n} a_{ij}(x,t)u_{x_i x_j} + \sum_{i=1}^{n} b_i(x,t)u_{x_i} + c(x,t)u - u_t = F(x,t,u), \qquad (5)$$
$$u(x,0) = f(x)$$

where

$$A|\xi|^2 \leq \sum a_{ij}\xi_i\xi_j \leq A^{-1}|\xi|^2, \quad \forall \xi \in \mathbb{R}^n$$

Assume the coefficients of (5) and initial value f(x) are sufficiently smooth and bounded. In [22], he proved when $c(x,0) \leq 0$ and there exists $\delta > 0$ such that

$$\sum_{i=1}^{n} b_i(x,t)x_i > \delta - \sum_{i=1}^{n} a_{ii}(x,t), \quad \forall(x,t) \in \mathbb{R}^n \times \mathbb{R}^+$$

Assume $F'_s(x,t,s) \geq 0$, and

$$|F(x,t,0)| \leq \frac{k(|x|)}{(t+1)^\alpha}, \quad \forall(x,t) \in \mathbb{R}^n \times \mathbb{R}^+$$

where $\alpha > 1$, $\lim_{s\to\infty} k(s) = 0$. If $\lim_{|x|\to\infty} f(x) = 0$, then the solution of (5) u(x,t) satisfies

$$\lim_{t\to\infty} u(x,t) = 0$$

for all $x \in \mathbb{R}^n$ uniformly.

In [24], he extended the above result to more general type. In [25], he extended the asymptotic property to Stefan problems.

The proof of asymptotic property involves simple and complicated methods of comparison of functions.

6   Prescribed boundary value problems of nonlinear parabolic equations

Consider the general second order nonlinear partial differential equation with two variables

$$F(x, y, u, p, q, r, s, t) = 0 \tag{6}$$

where $p = u_x$, $q = u_y$, $r = u_{xx}$, $s = u_{xy}$, $t = u_{yy}$. When $4F_r F_t - F_s^2 < 0(> 0)$ equation (6) are of hyperbolic or of elliptic type respectively, and they have proper prescribed boundary value problems for themselves. But what is the equation when it satisfies $4F_r F_t - F_s^2 = 0$? In this case the equation has double eigenvalues. To be parabolic type it needs some further conditions. In [26] [27], [28], some conditions of these types have been studied. Here we state the results of Wang [28] in which he proposed the conditions of these types and proved the existence and uniqueness of the first boundary value problems.

First we can consider $4F_r F_t - F_s^2 = 0$ is equavilent to the condition $(4F_r F_t - F_s^2)_{F=0} = 0$. That means

$$4F_r F_t - F_s^2 = AF,$$

where A is a smooth function of x, y, p, q, r, s, t. Next we introduce the eigenstrip $\{x(\alpha), y(\alpha), p(\alpha), q(\alpha), r(\alpha), s(\alpha), t(\alpha)\}$, where $x'^2 + y'^2 \neq 0$ and thw equation satisfies the strip conditions.

Assume that equation (6) is not degenerate, that means $F_r^2 + f_s^2 + F_t^2 \neq 0$ on the eigenstrip. We also assume that $F_r \neq 0$, then we can prove that the strip conditions are invariant under parameter transformation. The condition for r_y, s_y, t_y to be solvable is

$$\Delta \equiv \begin{vmatrix} -x'' + \frac{F_s}{F_r}x''^2 & -y'' + \frac{F_s}{F_r}x'^2 \\ x' & y' \end{vmatrix} \neq 0$$

Having the above conditions, we can solve the first boundary value problem with the initial value given on eigenstrip and boundary value given on two oblique curves near the strip.

By Schauder estimates, we can obtain a priori estimates of the solution for first boundary value problem and by Shauder fixed point theorem we can prove the existence of the solution.

As a special example in [28], the author discussed the general Monge-Ampere equation with two variables

$$H_r + 2Ks + Lt + M + rt - s^2 = 0,$$

where H, K, L, M are the functions of x, y, u, p and q. When the equation has double eigenvalues, i.e.,

$$K^2 - HL + M = 0$$

and $\Delta \neq 0$. We denote

$$\begin{aligned} D \equiv \ & (K_p x' - H_p y')L - [(K_q + L_p)x' - (H_q + K_p)y']K \\ & + (L_q x' - K_q y')H - (K_x + K_u p + L_y + L_u q)x' \\ & + (H_x + H_u p + K_y + K_u q)y' \end{aligned}$$

Assume that $D \neq 0$ on the eigenstrip, the condition is called the nondegenerate condition, then we can solve the first boundary value problem.

7   Strongly nonlinear variational problem

In [29] Ding and his colleagues considered the variational problem

$$I(u) = \inf_{v \in K} I(v), \quad I(v) = \int_{\Omega} E(x, v, v_x) dx,$$

where K is the set of functions v belonging to some Orlicz space and equal to $\phi(x)$ on the boundary. Suppose

$$|F(x, v, v_x)| \leq \mu(|v|)[L(x) + \phi(\beta|v|^p)],$$

where $\mu(t)$ is an increasing function on R^+, $L(x) \in L_1$, β, p are positive constants with $p > 1$ and $\phi(v)$ is any integral function. It is proved that there exist limited points of variational problem, moreover, the limited function u(x) satisfies the Euler equation of the variational. It extended the results of [30] to some fast increasing classes of functions.

In the proof of these results, they used Orlicz space as a tool and used the developed trace theorem of Orlicz space proposed by them.

In [31], Gu Y. G. and his colleagues considered the first boundary value problem of the nonlinear parabolic equation

$$\frac{\partial u}{\partial t} - \frac{d}{dx_i} a_i(x, t, u, u_x) + a(x, t, u, u_x) = 0,$$

where $a_i(x, t, u, s)$ and $a(x, t, u, s)$ are increasing fast in exponential growth of $s = (s_1, ..., s_n)$.

8   Subsonic flows in high dimensions

The steady irrotational subsonic flow of a perfect gas passing a given profile has been studied extensively. In [32], Bers proved the existence and uniqueness of plane subsonic flows around a given profile. R. Finn and D. Gilbarg [33] proved the existence and uniqueness in three dimensions provided the velocity is not larger than the velocity of sound. B. Ou [35] slightly modified the proof of [34] to show the existence and uniqueness in case of exterior forces. The main steps of Dong's proof, which has been simplified in [36], is sketched as follows.

The steady irrotational gas flow in n-dimensional space R^n can be described by the velocity potential $\phi(x)$. Denote $u_i = \frac{\partial \phi}{\partial x_i} (1 \leq i \leq n)$ the component of the velocity and $q = (\sum_{i=1}^{n} u_i^2)^{\frac{1}{2}}$ the velocity of gas. Then ϕ satisfies the equation

$$\sum_{i=1}^{n} \frac{\partial}{\partial x_i} (\rho \frac{\partial \phi}{\partial x_i}) = 0 \tag{7}$$

where $\rho = \rho(p)$ represents the density of gas. (7) can be written as

$$\sum_{i,j=1}^{n} a_{ij}(D\phi) \frac{\partial^2 \phi}{\partial x_i \partial x_j} = 0, \quad a_{ij} = \rho \delta_{ij} + \frac{\rho'}{\rho} \phi_{x_i} \phi_{x_j}, \quad \delta_{ij} = \begin{cases} 1, & i = j \\ 0, & i \neq j \end{cases}$$

The eigenvalues of (a_{ij}) are $\lambda_1 = \rho + \rho'q$, $\lambda_2 = ... = \lambda_n = \rho$. Assume that $\rho(0) = 1$, $\rho'(0) = 1$, and $\rho + \rho'q$ is decreasing in q. Denote q_c the zero point of $\rho + \rho'q$ (the critical velocity). The profile $\partial\Omega$ is a smooth closed surface. The region outside $\partial\Omega$ is denoted by Ω. Then the velocity potential $\phi(x)$ of subsonic flow around $\partial\Omega$ satisfies

$$\sum_{i,j=1}^{n} a_{ij}\frac{\partial^2\phi}{\partial x_i \partial x_j} = 0, \quad x \in \Omega,$$
$$\frac{\partial\phi}{\partial N}\Big|_{\partial\Omega} = 0, \quad \nabla\phi|_{x=\infty} = u^{\infty}, \tag{8}$$
$$\sup_{\overline{\Omega}}|\nabla\phi|(= Q) < Q_c,$$

where N is the outer normal of $\partial\Omega$, u^{∞} is a given constant vector, which is called the uniform incoming flow and can be taken as $(U, 0, ...,0)$. It is the main idea in [39] to deform the problem (8) with given U to the following auxiliary problem with given Q along a continuous (U, Q) curve:

$$\sum_{i,j=1}^{n} a_{ij}\frac{\partial^2}{\partial x_i \partial x_j} = 0,$$
$$\frac{\partial\phi}{\partial N}\Big|_{\partial\Omega} = 0, \quad \frac{\nabla\phi}{|\nabla\phi|}\Big|_{x=\infty} = (1,0,...,0), \tag{9}$$
$$\sup_{\overline{\Omega}}|\nabla\phi| = Q(< q_c).$$

The solution of the auxiliary problem (9) can be obtained by means of Leray-Schauder's fixed point theorem, which yields the unique solution of (8) (a continuous branch of solutions beginning at the zero solution). To be more precisely we introduce the following notations:

$$d_x = d(x,\partial\Omega), \quad d_{xy} = min(d_x, d_y),$$
$$M_{mk}(u) = \sup_{x\in\overline{\Omega}} max(1, d_x^{m+k})|D^k u|, \quad for \ u \in C^k(\overline{\Omega}),$$
$$M_{m,k+r}(u) = \sup_{x,y\in\Omega} max(1, d_{xy}^{m+k+2})\frac{|D^k u(x) - D^k u(y)|}{|x-y|^2},$$
$$for \ u \in C^{k+r}(\overline{\Omega})(0 < r \leq 1),$$
$$\|u\|_{mk} = \sum_{j=0}^{k} M_{mj}(u),$$
$$\|u\|_{m,k+r} = \|u\|_{mk} + M_{m,k+r}(u).$$

If $\|u\|_{m,k+r} < \infty$, then we say that $u \in C_{m,k+r}(\overline{\Omega})$.

Now we recall some well-known results about the following linear problem.

$$\mathcal{L}(u) = \sum b_{ij}(x)u_{x_i x_j} + \sum b_i(x)u_{x_i} + b(x)u = f(x),$$
$$\frac{\partial u}{\partial N}\Big|_{\partial\Omega} = u_0, \tag{10}$$
$$u|_{\infty} = 0.$$

Assume that there exist positive constants K and σ such that for any $\xi \in \mathbb{R}^n$

$$\sigma|\xi|^2 \leq \sum b_{ij}\xi_i\xi_j \leq \frac{1}{\sigma}|\xi|^2, \quad \|b_{ij}\|_{0,r} \leq K$$
$$\|b_i\|_{1,r} \leq K, \quad \|b\|_{2+r} \leq K$$
$$\|b_{ij}(x) - b_{ij}(\infty)\|_{r,0} \leq K, \quad \|b_i\|_{1+r,0} \leq K, \quad \|b\|_{2+r,0} \leq K,$$
$$f \in C_{n+2,0}(\overline{\Omega}) \cap C_{n,r}(\overline{\Omega}), \quad u_0 \in C^{1+r}(\partial\Omega).$$

Then the solution $u \in C^2(\Omega) \cap C^1(\overline{\Omega})$ of problem (10) exists and is unique. Moreover $u \in C^{2+r}(\overline{\Omega})$ and satisfies

$$M_{n-2,2+r}(u) \leq C_1(\|u_0\|_{C^{1+r}(\partial\Omega)} + \|f\|_{n,r} + \|f\|_{n+r,0})$$

The uniqueness of the solution and the priori estimate above can be obtained by the maximum principle of the exterior problem and Schauder estimate. By using the parameter extension method we can prove, step by step, the existence of the solutions, of the following equation

$$[(1 - \theta)\mathcal{L}_\infty + \theta \mathcal{L}]u = \theta f, \quad (0 \leq \theta \leq 1),$$
$$\frac{\partial u}{\partial N}\Big|_{\partial \Omega} = u_0, \quad u|_\infty = 0,$$

where $\mathcal{L}_\infty = \sum b_{ij}(\infty)\frac{\partial^2}{\partial x_i}\partial y_j$.

Since $Q < q_c$ in (9), it is not difficult to modify the equation (9) for $\nabla u \geq \frac{Q + q_c}{2}$, such that it becomes uniformly elliptic equation. Hence we assume without loss of generality, that

$$\frac{\sigma}{2}|\xi|^2 \leq \sum a_{ij}(\nabla\phi)\xi_i\xi_j \leq \frac{2}{\sigma}|\xi|^2, \quad \forall \xi \in \mathbf{R}^n, \quad \forall q \in [0, \infty).$$

Assume $\partial\Omega \in C^{2+r_0}$, taking $\phi(x) \in C^{1+r}(\overline{\Omega})$ such that $D\phi(\infty) = (U, 0, ..., 0)$, $[\phi(x) - U_{x_1}]_{x=\infty} = 0$, $|D\phi(x) - D\phi(\infty)| \leq C_1 d_x^{-r}$, $|D\phi(x) - D\phi(y)| \leq C_2 min(1, d_{xy}^{-r})|x - y|^r$, where $r = r(Q, \partial\Omega) \leq r_0$ (r will be determined later), the norm of $\phi(x)$ is defined by

$$\|\phi\| = |U| + \max_{\overline{\Omega}}|\phi - U| + infC_1 + infC_2$$

Such functions $\phi's$ form a Banach space E.

Given $\phi \in E$, consider the following

$$\sum a_{ij}(D\phi)\Psi_{x_i x_j} = 0,$$
$$\frac{\partial \Psi}{\partial N}\Big|_{\partial\Omega} = -cos(N, x_1)|_{\partial\Omega}, \quad \cdot$$
$$\Psi|_\infty = 0$$

The existence and uniqueness of Ψ is obtained by the linear theory mentioned above, and we have $\Psi \in C_{n-2,2+r}(\overline{\Omega})$. Hence $\max_{\overline{\Omega}}|\nabla(\Psi + x_1)| < \infty$.

Let $\Phi(x) = \dfrac{\theta Q[\Psi(x) + x_1]}{\max_{\overline{\Omega}}|\nabla(\Psi(x) + x_1)|}$, $0 \leq \theta \leq 1$. Then

$$\sum a_{ij}(\nabla\phi)\Phi_{x_i x_j} = 0,$$
$$\frac{\partial \Phi}{\partial N}\Big|_{\partial\Omega} = 0,$$
$$\frac{\nabla \Phi}{|\nabla \Phi|}\Big|_{\partial\Omega} = (1, 0, ..., 0)$$
$$\max_{\overline{\Omega}}|\nabla \Phi| = \theta Q$$

This implies that $\phi \to \Phi = T(\phi, \theta)$ is a mapping from E into E. The fixed point of this mapping is the solution of (9) with $\theta = 1$. In order to apply Leray-Schauder's fixed point theorem, the crucial step is to prove that the solutions of $\phi - T(\phi, \theta) = 0$ are bounded in E. Using the interior and nearly boundary (i.e. nearly $\partial\Omega$) estimates of the solution of (9) (with Q replaced by θQ), we can prove the bound of $infC_2$. We can also prove that $infC_1$ and $\phi \to U_{x_1}$ are bounded by means of the estimate in the neighborhood of ∞. The interior and nearly boundary estimates are classical results (see e.g. [30]). Since a_{ij} is approximately a constant in the neighborhood of ∞, the equation approximates $\Delta\phi = 0$. Therefore the estimate in the neighborhood of ∞ is obtained by technically using comparison functions. Now we can use Leray-Schauder's theorem to conclude that the equation (9) has at least one solution.

Now the uniqueness of solution of (8) and continuity of the curve (Q, U) yield the existence of solutions of (8). Furthermore we can also show that the solutions continuously depend on the magnitude and directions of uniform incoming flow.

References

[1] Visik M.I., On strongly elliptic systems of differential equations, Math. Sbornik, 29(1951), 613-676.

[2] Ding X.Q., Wang K.T., Ma, Zhang T. and Sun J.L., Definition of ellipticity of a system of second order differntial equations with constant coefficients, Acta Math. Sinica, 10(1960), 276-287.

[3] Hua L.G., Wu. Z.Q. and Lin. W., On the uniqueness of the solution of the Dirichlet problem of an elliptic systems of differential equations with constant coefficients, Acta Math. Sinica, 15(1965), 242-248.

[4] Hua L.G., Wu Z.Q. and Lin W., A system of linear partial differential equations of the second order with constant coefficients, two variables and two unknown functions, Academic Press, (1979).

[5 & 6] Li D.Q., Zheng S.M., Tan Y.J., Shang H.J., Gao R.X. and Sheng W.X., Boundary value problems with equal value surface boundary conditions for selfadjoint elliptic differential equations, Fudan Journal, (1976),61-71, 136-145.

[7] Li D.Q., Chen S.X., Zheng S.M. and Tan Y.J., On boundary value problems of the selfadjoint elliptic equation with complementary boundary conditions, Fudan Journal, (1978), 49-60.

[8] Chen S.X., On boundary value problems of the second order partial differential equation with complementary boundary conditions, Fudan Journal, (1979), 37-44.

[9] Agman S., Douglis A. and Nirenberg L., Estimates near the boundary for solutions of elliptic partial differential equations satisfying general boundary conditions, Comm. Pure Applied. Math., 7(1959), 623-727.

[10] Friedman A., Interior estimates for second order parabolic equations and their applications, J. Math. Mech., 7(1958), 393-417.

[11] Friedman A., Boundary estimates for second order parabolic equations and their applications, J. Math. Mech., 7(1958), 771-809.

[12] Teng Z.H., A prior estimate of solutions for parabolic equations with general boundary, Advances in Math.(China), 8(1965), 352-386.

[13] Wang R.H., On the Schauder type theory about the general parabolic boundary value problems, Acta Sientiarum Naturalium Universitis Jilinensis, (1964), 35-64.

[14] Dong G.C., On the Riemann-Hilbert problem in multiply-connected domain, Sci. Record, 2(1958), 131-134.

[15] Li Z. and Wen G.C., On the Riemann-Hilbert boundary value problem of elliptic systems of nonlinear partial differential equations of the first order, Acta Math. Sinica, 15(1965), 599-613.

[16] Li Z. abd Wen G.C., The number of solutions of generalized Riemann- Hilbert boundary value problem, Acta Math. Sinica, 15(1965), 765-774.

[17] Wen G.C., The Riemann-Hilbart problem for nonlinear elliptic systems of first order on the plane, Acta Math. Sinica, 23(1980), 244-255.

[18] Hou Z.Y., Boundary value problem for second order elliptic systems containing a singular line, Fudan Journal (Natural Science), (1979), 50-62.

[19] Lu W.D. and Wang J.H., The maximum principles for some semi-linear elliptic equations of fourth order and their applications, Journal of Sichuan University, 22(1981), 33-45.

[20] Lu L.J., A uniqueness theorem for the first boundary value peoblems of the linear elliptic and parabolic equations with discontinuous boundary values, Acta Math. Sinica, 15(1965), 372-385.

[21] Shen Y.T., The Dirichlet problem for degenerate or singular elliptic equations of higher order, Journal of the China University of Science and Technology, 10(1980), 15-25.

[22] Gu L.K, Asymptotic behavior of the solution of the Cauchy problem for a parabolic equation, Acta Math. Sinica, 12(1962), 284-292.

[23] Gu L.K., Asymptotic behavior of solution for a quasilinear parabolic equation, Advances in Math. (China), 6(1963), 272-278.

[24] Shi X.L., Li M.D. and Qin Y.C., On asymptotic behavior of solutions for Cauchy problem of nonlinear parabolic equations, Acta Math. Sinica, 24(1981), 451-463.

[25] Gu L.K., The asymptotic behavior of the solution of the multiphase Stefan problem, Acta Math. Sinica, 23(1980), 203-214.

[26] Wang G.Y., Quasilinear parabolic equations, Acta Math. Sinica, 14(1964), 494-502.

[27] Wang G.Y., Chen D.D. and Mei D.W., On Monge-Ampére equation of parabolic type, Acta Math. Sinica, 14(1964), 78-92.

[28] Wang G.Y. and Mai M.C., Nonlinear parabolic equations, Acta Math. Sinica, 16(1966), 283-299.

[29] Ding X.Q., Luo P.Z., Gu Y.G. and Fang F.Z., Calculus of variations with strong nonlinearity, Scientia Sinica, 23(1980), 945-955 (Chinese and English).

[30] Ladyzenskaja O.A. and Uralceva N.N., Linear and quasilinear equations of elliptic type "Nauka" Moscow, (1965).

[31] Gu Y.G., Luo P.Z. and Ding X.Q., Generalized solutions of strongly nonlinear parabolic equations, Scientia Sinica (Series A), V.26, No.11 (1983), 1129-1143.

[32] Bers L., Existence and uniqueness of subsonic flows past a given profile, Comm. Pure Appl. Math., 7(1954), 441-504.

[33] Finn R. and Gilbarg D., Three-dimensional subsonic flows and asymptotic estimates for elliptic partial differential equations, Acta Math., 98(1957), 265-296.

[34] Dong G.C., Three-dimensional subsonic flows and their boundary value problems extended to higher dimensions, Journal of Chekiang University, (1979), 33-63.

[35] Ou B., On the existence, uniqueness and continuous dependence of solution for three dimensional subsonic flows with "source", preprint.

[36] Dong G.C., Nonlinear partial differential equations of second order, Qinghua University Press, (1988), Translations of Mathematical Monographs, Vol.95, AMS(1991).

Viscosity Solutions of Fully Nonlinear Elliptic and Parabolic Equations
— Some works of recent development on the basis of a-priori estimation[†]

Guangcang Dong and Baojun Bian

Department of Mathematics, Zhejiang University, Hangzhou

In this expository paper we wish to describe some recent progress in viscosity solutions of fully nonlinear second order elliptic and parabolic equations. We shall concentrate on the Dirichlet problem of elliptic equations and the first boundary value problem of parabolic equations.

The Dirichlet problem of elliptic equations is as follows

$$F(x, u, Du, D^2u) = 0, \quad \text{in } \Omega \tag{1}$$
$$u = \phi, \quad \text{on } \partial\Omega \tag{2}$$

where Ω is a bounded domain in $\mathbb{R}^n$, and

$$f(x, z, p, r) \leq F(x, z, p, r + s), \quad \forall s \geq 0 \tag{3}$$

for all $(x, z, p, r) \in \Omega \times \mathbb{R} \times \mathbb{R}^n \times \$^n$, here $\$^n$ is the set of symmetric $n \times n$ matrices and is equiped with its usual order.

The condition (3) is called the degenerate elliptic condition of F.
The first boundary value problem of parabolic equations is as follows

$$F(x, t, u, Du, D^2u) - u_t = 0, \quad \text{in } Q \tag{4}$$
$$u = \varphi, \quad \text{on } \partial^0 Q \tag{5}$$

where $Q = \Omega \times (0, T]$, $\partial^0 Q = (\Omega \times \{t = 0\}) \cup (\partial\Omega \times [0, T])$ and

$$f(x, t, z, p, r) \leq F(x, t, z, p, r + s), \quad \forall s \geq 0 \tag{6}$$

for all $(x, t, z, p, r) \in Q \times \mathbb{R} \times \mathbb{R}^n \times \$^n$.

[†] 1991 Mathematics Subject Classification: 35J25, 35J65, 35K60
[†] Supported by NNSF of China and Foundation of Science of Zhejiang Province

C. Gu et al. (eds.), Partial Differential Equations in China, 42–49.

Both equations arise in the theory of Monge-Ampere equation on the convex surface and Bellman equation on the controlled diffusion process. Many mathematicians contributed to the theory of the above equations and their results are described in books [1], [2], [3], and for other boundary value problem in papers [4], [5], [6].

For the existence result of classical solutions of these equations in uniformly nondegenerate case, three sets of conditions are needed.

i) Some conditions impose on F for the estimation of the bounds of $||u||_{C(\overline{\Omega})}$, $||u||_{C^\alpha(\overline{\Omega})}$ (some $\alpha \in (0,1)$) for (1) and (2).

ii) Some conditions impose on the first derivatives of F, i.e., F_x, F_z, F_p, F_r for the estimation of the bounds of $||u||_{C^1(\overline{\Omega})}$, $||u||_{C^{1+\alpha}(\overline{\Omega})}$. These are called the natural structure conditions.

iii) Condition impose on the second derivatives of F for the estimation of the bounds of $||u||_{C^2(\overline{\Omega})}$, $||u||_{C^{2+\alpha}(\overline{\Omega})}$. It is called the convex type condition.

Similar conditions are needed for parabolic case (4) and (5).

The convex type condition imposing on F for the existence of solution of (1), (2) or (4), (5) is quite unnatural and serious. In order to remove it, we must impose the concept of generalized solutions.

Since the PDE (1) and (4) cannot be put into divergence form, we are not able to define a weak solution by means of formal integration by parts of derivatives onto a smooth test function. We will adopt the notion of viscosity solution.

The notion of viscosity solution was introduced by M. G. Crandall and P. L. Lions in [7] first for Hamilton-Jacobi equations. It is also valid for second order PDE (1) and (4).

Definition The upper semicontinuous function u(x), $x \in \Omega$, is called the viscosity subsolution of (1) if for all $\varphi \in C^2(\Omega)$, on the local maximum point $x_0 \in \Omega$ of $u - \varphi$ we have

$$F(x, u, D\varphi, D^2\varphi)|_{x=x_0} \geq 0 \qquad (7)$$

The lower semicontinuous function u(x), $x \in \Omega$, is called the viscosity supersolution of (1) if for all $\varphi \in C^2(\Omega)$, on the local minimum point $x_0 \in \Omega$ of $u - \varphi$ we have

$$F(x, u, D\varphi, D^2\varphi)|_{x=x_0} \leq 0 \qquad (8)$$

Then, $u \in C(\Omega)$ is said to be a viscosity solution of (1) if it is a viscosity subsolution and a viscosity supersolution.

It is clear that the classical solution of (1) (or (4)) is the viscosity solution of (1) (or (4)).

The definition of viscosity solution transfers the derivatives of solution to the smooth test function. So that it is a better generalized solution fitted by fully nonlinear equations.

The great progress in studying the viscosity solutions was made by R. Jensen [8], H. Ishii and P.L. Lions [9]. In [8] the author has proved the uniqueness of viscosity solutions of (1) and (2) without the convex condition on F, where F is independent of x. B.J. Bian [10] extended the result of [8] to more general case of F.

In [9] authors established the fundamental inequality for viscosity solutions of (1), and got many results, such as the comparison, uniqueness, existence and regularity properties of viscosity solutions of (1).

The first step for establishing the fundamental inequality of PDE (1) is the construction of the appropriate smooth approximation of bounded upper and lower semicontinuous functions u(x) and v(x) by

$$u^{\varepsilon}(x) = \sup_{y \in \Omega}\{u(y) - \frac{|x-y|^2}{\varepsilon}\}, \quad x \in \Omega_{\varepsilon} \tag{9}$$

$$v_{\varepsilon}(x) = \inf_{y \in \Omega}\{u(y) + \frac{|x-y|^2}{\varepsilon}\}, \quad x \in \Omega_{\varepsilon} \tag{10}$$

where $\Omega_{\varepsilon} = \{x \in \Omega | d(x, \partial\Omega) > \varepsilon\}$. The functions u^{ε} and v_{ε} have a series of properties such as, both are Lipschitz continuous in x (with Lipschitz constant depends on ε).

One extension we adopted of (9) and (10) to parabolic case is: when u(x,t) and v(x,t) are bounded upper semi-continuous and lower semi-continuous respectively, define

$$u^{\varepsilon}(x,t) = \sup_{(y,\tau) \in Q \cap \{\tau \le t\}}\{u(y,\tau) - \frac{|x-y|^2 + t - \tau}{\varepsilon}\} \tag{11}$$

$$v_{\varepsilon}(x,t) = \inf_{(y,\tau) \in Q \cap \{\tau \le t\}}\{v(y,\tau) + \frac{|x-y|^2 + t - \tau}{\varepsilon}\} \tag{12}$$

for $(x,t) \in Q_{\varepsilon} = \{(x,t) \in Q | d((x,t), \partial^{\alpha}Q > \varepsilon\}$. Formula (11) and (12) are defined in [11] of us, and thus we established the comparison and uniqueness results of viscosity solutions of (4), (5).

We proved that $u^{\varepsilon}(or v_{\varepsilon})$ is continuous in t when u (or v) is upper (or lower) semicontinuous viscosity subsolution (or supersolution). On this basis we extended the fundamental inequality of viscosity solutions of (1) to (4) as follows.

Theorem Let F(x,t,z,p,r) satisfy (6) and

$$F(x,t,z_1,p,r) \ge F(x,t,z_2,p,r) \qquad \forall z_1 \le z_2$$

and u(x,t), v(x,t) be the bounded subsolution and supersolution of (4) respectively. Let $\varphi(x,y,t) \in C^{2,2,1}(\Omega \times \Omega \times (0,T])$ and $w(x,y,t) = u^{\varepsilon}(x,t) - v_{\varepsilon}(y,t)$ for $(x,t), (y,t) \in Q_{\varepsilon}$, where u^{ε} and v_{ε} are defined by formula (7) and (8). Assume that $w - \varphi$ achieves its maximum over $\overline{\Omega} \times \overline{\Omega} \times [\varepsilon^2, T]$ at the point $(\bar{x}, \bar{y}, \bar{t}) \in \Omega_{\varepsilon} \times \Omega_{\varepsilon} \times (\varepsilon^2, T]$. Then there exist matrices $X, Y \in \$^n$ such that

$$F_{\varepsilon}(\bar{x}, \bar{t}, u^{\varepsilon}(\bar{x}, \bar{t}), D_x\varphi(\bar{x}, \bar{y}, \bar{t}).X) - F^{\varepsilon}(\bar{y}, \bar{t}, v_{\varepsilon}(\bar{y}, \bar{t}), -D_y\varphi(\bar{x}, \bar{y}, \bar{t}), -Y)$$
$$\ge \varphi_t(\bar{x}, \bar{y}, \bar{t}).$$

and

$$-\frac{2}{\varepsilon}I \le \left(\begin{array}{cc} X & 0 \\ 0 & y \end{array} \right) \le D^2_{xy}\varphi(\bar{x}, \bar{y}, \bar{t}),$$

where

$$F_\varepsilon(x, t, z, p, r) = sup\{F(y, s, z, p, r)|(y, s) \in Q_{C_v, \varepsilon_{1/2}}(x, t)\}$$
$$F^\varepsilon(x, t, z, p, r) = inf\{F(y, s, z, p, r)|(y, s) \in Q_{C_v, \varepsilon_{1/2}}(x, t)\}$$
$$C_0 = max\{2\sup_Q |u|, \; 2\sup_Q |v|\}.$$

We got the comparison and uniqueness results of viscosity solutions of (4). The comparison theorem for (4) is that for any subsolution u and supersolution v of (4) satisfying (6), (9) and u, v $\in C(\partial^\alpha Q)$, we have

$$\sup_Q(u - v) = \sup_{\partial^\alpha Q}(u - v)$$

On the basis of comparison result, we completed in [12] the Perron's construction of solution of (4) and (5). The method of proof is similar to that in [13].

However the construction of solution to the problem (1), (2) and (4), (5) by Perron's method based on the existence of a pair of continuous viscosity subsolution and supersolution which satisfy the boundary condition (2) or (5). But in general, unlike in the linear case, it is quite difficult to find such a pair of subsolution and supersolution for the fully nonlinear PDE. The difficulty is nearly equal to finding a viscosity solution. In other words, the construction of viscosity solution to (1), (2) or (4), (5) by Perron's method is not an actual proof for the existence of solution. So that to prove the existence of viscosity solutions to (1), (2) or (4), (5) by other method is needed. In this direction we are successful in fully nonlinear equation under natural structure conditions. In [14] the existence of Lipschitz viscosity solution to (1) and (2) was proved. In [15] the result of [14] was extended to the parabolic case (4) and (5).

The conditions and steps of proof given in [15] are as follows.

Theorem Let F(x,t,z,p,r) satisfy the uniformly elliptic condition

$$\lambda I \le F_r(x, t, z, p, r) \le \lambda^{-1}I \tag{13}$$

where $\lambda > 0$ and I is the unit matrix. Conditions for estimation of $\|u\|_{C^0}$

$$F_x(x, t, z, p, r) \le const. \tag{14}$$
$$|F(x, t, 0, 0, 0)| \le const. \tag{15}$$

Natural structure condition.

$$|F| + (1 + |p|)|F_p| + |F_x| + \frac{1}{1 + |p|}|F_p| \le G(sup|z|)(1 + |p|^2 + |r|). \tag{16}$$

Then a viscosity solution $u \in W^{1,0}_\infty(Q) \cap C^{\alpha, \frac{\alpha}{2}}(\overline{Q})$ of (4) and (5) exists.

Sketch of the proof.

We use the idea of accretive operator and m-accretive operator, their definition and properties see [16].

Without loss of generality we assume u=0 on $\partial^\alpha Q$. Write equation (4) into

$$\frac{\lambda}{2}\Delta u + G(x,t,u,Du,D^2u) - u_t = 0.$$

Since a convex function of (z,p,r) can be written as the envelop of its tangent planes, and any continuous function of (z,p,r) can be written as the envelop of convex functions. Hence G can be written as follows

$$G(x,t,u,Du,D^2u) = \sup_{(z_1,p_1,r_1)} \inf_{(z_2,p_2,r_2)} L_{(x,t,z_1,p_1,r_1,z_2,p_2,r_2)}u,$$

where $L(x,t,z_1,p_1,r_1,z_2,p_2,r_2)$ are linear operators. Approximate the homogeneous part of L, $L_0 = L - L|_{u=D_zu=D_z^2u=0}$ by

$$A_\eta = \frac{1}{\eta}[I - (I + \eta^2\frac{\partial}{\partial t} + \eta L_0)^{-1}].$$

Then A_η is a Lipschitz continuous accretive operator defined on $C(\overline{\Omega})$ for all $(z_i,p_i,r_i) \in \mathbb{R} \times \mathbb{R}^n \times \$^n$ (i=1,2).

Take a cut-off function $\beta = \beta_\epsilon(s)(s \in \mathbb{R})$ such that $\beta(s) = 1$, $|s| \leq \frac{1}{\epsilon} - 1$; $\beta(s) = 0$, $|s| \geq \frac{1}{\epsilon}$; $\beta'(s) \leq const.$ Let

$$B_{\epsilon,\eta}(u) = \beta_\epsilon[\sup_{\sum_{i=1}^2(|z|_i^2+|r_i|^2+|r_i|^2)\leq C_\epsilon} \inf A_\eta(u,D_zu,D_z^2u) + L - L_0]$$

Then $B_{\epsilon,\eta}$ is an everywhere defined, Lipschitz continuous, accretive operator in $C(\overline{Q})$,. Hence $\frac{\lambda}{2}\Delta + B_{\epsilon,\eta} - \frac{\partial}{\partial t}$ is a m-accretive operator and a solution $u = u_{\epsilon,\eta} \in W_p^{2,1}(Q) \cap W_p^{1,1}(Q)$ of

$$\frac{\lambda}{2}\Delta u - \eta u + B_{\epsilon,\eta} - \frac{\partial}{\partial t}u = 0$$

exists for all $p \in [1,\infty)$.

Let $\eta \to 0$ we get a function u_ϵ satisfying $u_\epsilon \in W_p^{2,1}(Q) \cap W_p^{1,1}(Q)$, $u_\epsilon = 0$ on $\partial^\alpha Q$, and

$$\frac{\lambda}{2}\Delta u_\epsilon + \beta_\epsilon(G(x,t,Du_\epsilon,D^2u_\epsilon)) - \frac{\partial}{\partial t}u_\epsilon = 0$$

Estimate the bounds of $\|u_\epsilon\|_C^0$, $\|u_\epsilon\|_{C^{\alpha,\frac{\alpha}{2}}}$, $\|u_\epsilon\|_{C^1}$, independent of ϵ similar to [3]. But since $D_x^3u_\epsilon$ do not exist, we should use the finite difference quotient $U(\xi,h) = \frac{u_\epsilon(x+h\xi,t)-u_\epsilon(x,t)}{h}$, $|\xi| = 1$, instead of differentiation with respect to x. The estimation is successful by adding auxiliary variable ξ in the region $(x,t,\xi) \in Q \times \{|\xi| < 1\}$.

After these estimations, we let $\epsilon \to 0$, a viscosity solutions of elliptic and parabolic equations, we start out to discuss the regularity of viscosity solutions. It is well known that the regularity is very important for weak solutions. Some authors have studied this problem and have obtained a series of results.

Notice that the solutions constructed by accretive operator method as in [14] and [15] are in $W^{1,\infty}(\Omega)$ (or $W_\infty^{1,0}(Q_T) \cap C^{\alpha,\frac{\alpha}{2}}(\bar{Q}_T)$) in parabolic case for some $0 < \alpha < 1$, and that constructed by Perron's method are only in $C(\bar{\Omega})$ (or $C(\bar{Q}_T)$). We want to obtain better regularity.

In [9] H. Ishii and P.L. Lions considered the regularity problem for elliptic equations. They proved the Hölder continuity for fully nonlinear equations under the conditions which are as same as that used in studying the uniqueness theorem. The semiconcavity was given for the elliptic type Bellman equations. These regularity results were derived by means of comparison principle which was established in [9].

The regularity of the derivatives of viscosity solutions was first investigated by L. A. Caffarelli [17]. By using the perturbation technique he considered a class of fully nonlinear elliptic equations

$$F(D^2 u, x) = f(x) \qquad (17)$$

and proved the following theorem: if the solutions of the freezing equations of (17)

$$F(D^2 \omega, 0) = 0 \qquad (18)$$

exist and satisfy the priori $C^{1,\bar{\alpha}}$ interiori estimates for some $0 < \bar{\alpha} < 1$, then for any $0 < \alpha\bar{\alpha}$, the bounded solutions of (17) is in $C^{1,\alpha}$ provided f satisfy some integrability conditions and the oscillation of F is small enough.

Pertubation technique allows one to obtain the regularity of the derivatives without differentiating the equations.

The proof in Caffarelli's regularity theorem contains two main ingredients. The first is the establishment of the Aleksandrov-Bakel'man -Pucci type maximum principle for viscosity solutions. It plays an important role in the regularity theory. The second is to approximate the solutions of (17) by linear functions with arise from the "good" solutions of the freezing equation by means of the compactness argument.

In [17] L.A. Caffarelli also got the $W^{2,p}$ and $C^{2,\alpha}$ estimates by using the same argument under the assumptions of good estimates for freezing equations.

A natural problem is to study the regularity of viscosity solutions for elliptic and parabolic equations involving the lower order term. In this direction, L. Wang [18], G. Dong and B. Bian [19] generalized Caffarelli's regularity thoery to equations with more general form.

L. Wang extended the regularity theory in [15] to parabolic case with some improvement. By using the Aleksandrov maximum principle and the method of compactness he obtained Harnack, $C^{1,\alpha}$, $C^{2,\alpha}$ and $W^{2,p}$ estimates for parabolic equations (4) under the assumption of good estimates for the equation in which the lower order terms have been frozen. As in [17], the main idea of [18] is to understand the partial derivatives of functions by their polynomial approximations. However the regularity results in [18] were limited to case when the L^∞ norm of solutions is small enough or F grows linearly on Du.

In [19] G. Dong and B. Bian adapted and extended this perturbation approach and also applied it to a class of elliptic and parabolic equations which satisfy the natural construction conditions. They proved that if the natural construction conditions are satisfied, boundary $\partial\Omega$ (or $\partial_* Q_T$) and boundary data have appropriate smoothness and

$$|F_l(x, z, p, r)| \leq C + \mu|r|$$
$$(|F_l(x, t, z, p, r) \leq C + \mu|r| \quad \text{in parabolic case})$$

for $\mu > 0$ small enough, then the viscosity solutions to (1) and (4) have global $C^{1,\alpha}$ regularity for some $0 < \alpha < 1$.

The first step of proof is to approximate voscosity solutions by $W^{2,p}$ (p large enough) strong solutions according to accretive operator method. Then the $C^{1,\alpha}$ uniform estimates for $W^{2,p}$ strong solutons of equations which do not involve Du were derived with the aid of the difference quotient argument. Finally the desired results follow by applying the perturbation technique.

We shall discuss briefly the viscosity solutions of obstacle problem for fully non-linear elliptic equations. In [20] the author proved the existence and uniqueness results of viscosity solutions for the obstacle problem of fully nonlinear elliptic equations under the natural construction conditions. We do not assume that F is convex. The technique we used is as same as that in [14].

References

[1] D. Gilbarg and N.S. Tudinger, Elliptic partial differential equations of second order, 2nd Edition, Spronger-Verlag, New York, 1983.

[2] N.V. Krylov, Nonlinear elliptic and parabolic equations of second order, D. Reidel Publishing Co., Boston, 1987.

[3] G.C. Dong, Second order nonlinear partial diferential equations, Tsinghua University Press, 1988.

[4] G. M. Lieberman, The nonlinear oblique derivative problem for quasilinear elliptic equations, Nonlinear Anal. 8(1984), 49-66.

[5] G.M. Lieberman and N.S. Trudinger, Nonlinear oblique boundary value problems for nonlinear elliptic equations, Trans. Amer. Soc. 295(1986), 505-546.

[6] G. C. Dong, Initialand nonlinear oblique boundary value problems for fully non-linear parabolic equations, J. Partial Diff. Equa. Vol. 1, No.2 (1988), 12-42.

[7] M.G. Crandall and P.L. Lions, Viscosity solutions of Hamilton-Jacobi equations, Trans. Amer. Math. Soc. 277(1983), 1-42.

[8] R. Jensen, The maximum principle for viscosity solutions of fully nonlinear second order partial differential equations, Arch. Rat. Mech. Aral. 101(1988), 1-27.

[9] H. Ishii and P.L. Lions, Viscosity solutions of fully nonlinear seconf order elliptic partial differential equations, J. Diff. Equa. 83(1990), 26-78.

[10] B. J. Bian, The uniqueness of viscosity solutions of the second order fully non-linear elliptic equations, J. Partial Diff. Equa., Vol. 1, No.1(1988), 77-84.

[11] G.C. Dong and B.J. Bian, Uniqueness of viscosity solutions of fully nonlinear second order parabolic PDE's, Chin. Ann. of Math., 11B: 2(1990), 156-170.

[12] G.C. Dong and B.J. Bian, Viscosity solutions of fully nonlinear second order parabolic PDE's, Preprint.

[13] H. Ishii,   Perron's method for Hamilton-Jacobi equations, Duke Math. J. 55(1987), 369-384.

[14] B. J. Bian, Existence of viscosity solutions of second order fully nonlinear elliptic equations, to appear.

[15] G.C. Dong, Viscosity solutions of second order fully nonlinear parabolic equations, to appear in Comm. P.D.E..

[16] L.C. Evans, On solving certain nonlinear differential equations by accretive operator methods, Isreal J. Math. 36(1980), 225-247.

[17] L. A. Cffarelli, Interior a priori estimates for solutions of fully nonlinear equations, Ann. Math., 130(1989), 189-213.

[18] L. Wang, On the regularity theory of fully nonlinear parabolic equations, to appear.

[19] G.C. Dong and B.J. Bian, The regularity of fully nonlinear equations, to appear in Scientia Sinica.

[20] B.J. Bian, Viscosity solutions of the obstacle problem for fully nonlinear elliptic equations, Preprint.

Some Developments of the Theory of Mixed PDEs [†]

Chaohao Gu [††] and Jiaxing Hong[†††]

Institute of Mathematics, Fudan University, Shanghai

The theory of boundary value problems for mixed equations was initiated in 1923 by F.G. Tricomi [T] and attracted much attention of many mathematicians by F. Frankl's work [F] which revealed the closed connection between this theory and the transonic flow dynamics. However, the study of mixed equations is a quite difficult field. As [P4] stated, " to form a single comprehensive theory for mixed equation in $I\!R^2$, at this time, appears formidable, and the development in three and more dimensions and for equations of order other than two is even more remote ". One can see the difficult degree from this comment. The purpose of the present paper is to introduce a part of Chinese mathematicians' work on mixed equations. The materials extracted here are mainly based on [DC] and [GH], particularly on the latter. In view of the limitation of the space of the present paper, all the results on the degenerate elliptic and the degenerate hyperbolic equations, and applications to Differential Geometry and Mathematical Physics are precluded. It should be emphasized that the choice of the material depends much on the authors' taste and the present paper is by no means an exhaustive survey of the researches in China of this field.

1   Tricomi problems

Chinese mathematicians began to study mixed type equations with Tricomi's pioneering work and in a large degree, were inspired by the connection of mixed equations to the theory of transonic flow dynamics. Most of the results in early fifties are concerning the uniqueness of Chaplygin's equation, which is a generalization of Tricomi equation, namely

$$K(y)u_{xx} + u_{yy} = 0 \quad in \quad \Omega \tag{1.1}$$

---

† 1991 Mathematics Subject Classification: Primary 35M05, 35M10, Secondary 35A27
†† Supported by the NNSF of China and the Education Commission DF of China.
††† Supported by the NNSF of China.

C. Gu et al. (eds.), Partial Differential Equations in China, 50–66.

where K is a C^2 function and $K(0)=0$, $yK > 0$, $y \neq 0$ and $K' > 0$ in the region consideration, and Ω is enclosed by three curves: AB, AC and BC. Here AB is a Jordan curve in the upper plane with the endpoints $A(x_1,0)$, $B(x_2,0)$, AC and BC are two characteristic curves in the below plane, satisfying respectively

$$dy = -\sqrt{K}\,dx \text{ and } dy = \sqrt{K}\,dx$$

The boundary conditions prescribed are as follows,

$$u = \phi \quad on \quad AB \cup BC \tag{1.2}$$

for some sufficiently smooth function ϕ. This is so called Tricomi problem.

In [WD] using the so called " a.b.c " method, Wu, S.M. and Ding, S.S. gave a different type of uniqueness theorem from [P1] to generalize the result in [F]. This is the first result on mixed equation in China. Afterwards various further results on the uniqueness of solutions were also obtained by Wang, K.Y. [W], Dong, G.C.[D1] with the aid of the "a.b.c" method like the energy integral ways. Note that in the L^2 sense, the uniqueness of strong solutions to a boundary value problem is equivalent to the existence of weak solutions of the conjugate problem. Therefore these results on uniqueness mentioned above indeed also made the contribution to the existence of weak solutions.

Let a graph in $I\!\!R^3$ be defined by a C^2 function $z=f(x,y)$ over a domain $\Omega \subset I\!\!R^2$. E.H. Bekua pointed out that the vertical component u of the infinitesimal isometric deformation vector of this graph satisfies

$$u_{xx}f_{xx} - 2u_{xy}f_{xy} + u_{yy}f_{yy} = 0 \quad in \quad \Omega \tag{1.3}$$

(1.3) is elliptic or hyperbolic if $\Delta = f_{xx}f_{yy} - f_{xy}^2 > 0$ or < 0 and hence it is the equation of mixed type if the curvature K of this graph changes its sign. Sun, H.S. studied the well-posedness of Tricomi problem of (1.3). Denote by $\Omega_+ = \Omega \cap \{(x,y)|\Delta > 0\}$ and by $\Omega_- = \Omega \cap \{(x,y)|\Delta < 0\}$, Assume that $\partial\Omega_+ \cap \partial\Omega$ is a Jordan curve AB and $\partial\Omega_- \cap \partial\Omega$ composed of two characteristic curves AC and BC. Under some mild restrictions on the function f and the geometry of the domain [SU1] presented the uniqueness of the following Tricomi problem

$$u = \phi \quad on \quad BA \cup AC \tag{1.4}$$

Thus a kind of rigidity theorem of the surfaces was obtained. Recently, some result on the existence of strong solution of the above Tricomi problem has been obtained in [SU2] for the case of the nonhomogeneous equation and the homogeous boundary condition.

In study of the transformation on the real projective plane, preserving the unit circle Hua, L.K. [HU1-2] derived the following mixed type equation

$$(\delta_{ij} - x_i x_j)\phi_{\cdot ij} - 2x_i\phi_{\cdot i} = 0 \quad in \quad I\!\!R^2 \tag{1.5}$$

Here and later the summation convention is used and $\phi_{.i}$ and $\phi_{.ij}$ denote the derivatives of ϕ

$$\phi_{.i} = \frac{\partial \phi}{\partial x_i} \quad and \quad \phi_{.ij} = \frac{\partial^2 \phi}{\partial x_i \partial x_j}$$

(1.5) is the Busemann equation derived by Busemann from the aerodynamics. This equation is hyperbolic outside the unit circle and elliptic inside the unit circle and moreover, can be tranformed to the wave equation and Laplace equation respectively by the characteristic coordinates. Using the method of separate variables, Hua [HU1–2] gave an explicit formula of the solutions to (1.5) in the class of the functions which is, in the region considered, continuous and twice continuously differentiable except the unit circle, subject to such a matching condition on the unit circle

$$\lim_{\rho \to 1-0} \frac{\phi(\rho,\theta) - \phi(1,\theta)}{\sqrt{|1-\rho|^2}} = \lim_{\rho \to 1+0} \frac{\phi(1,\theta) - \phi(\rho,\theta)}{\sqrt{|1-\rho^2|}}$$

With the aid of this explicit formula the uniqueness and existence of solutions to a Tricomi like problem were solved. Later, some extensions were made by Ji, X.H. and Chen, D.Q. [JC].

2 Second order equations in higher dimensional spaces

In the early sixties very little was known about the mixed equations in higher dimensional spaces except some theorems of uniqueness on some regions in $\mathbb{R}^3$ (see [P3]). It was K.O., Friedrichs who noted the fact that the energy integral method is applicable to all the three kinds of the classical (elliptic, hyperbolic and parabolic) equations and, in 1958, initiated the theory of the positive symmetric system [FK] to open up a complete new way for the study of equations of mixed type. Indeed, he studied some boundary problems in $\mathbb{R}^2$. But in using Friedrichs/ theory there were some crucial difficulties. namely, as [JA] pointed out, no general methods to reduce an arbitrary boundary value problem to an admissible problem for a certain positive symmetric system. Also at that time no one knew if any mixed equation more complicated than Tricomi or Chaplygin equation could be reduced to a positive symmetric system. However, the first author of the present paper considered the solutions of higher differntiability to the positive symmetric system, and extended Friedrichs/ theory to quasilinear case. Particularly, he found a general method for the reduction of mixed equations to a positive symmetric system and gave some results on existence of strong solutions, including classical solutions, to a class of boundary value problems for the generalized Busemann equation in higher dimensional spaces (see [G2]). To author/s knowledge, this is the first result on the existence for the mixed equation in higher dimensional space. Furthermore, with the aid of these developments, a large class of linear as well as quasilinear mixed equations in higher dimensional spaces were studied and some interesting and surprising phenomena were discovered. For details, see [GU1-8]. The present section intends to introduce the general method mentioned above.

At first he considered the generalized Busemann equation with a parameter a

$$(\delta_{ij} - x_i x_j)\phi_{.ij} + 2ax_i\phi_{.i} - a(a+1)\phi = f \quad in \quad \Omega \subset I\!R^n \qquad (2.1)$$

(2.1) is coincide with (1.5) if a=-1 and n=2. So we call it the generalized Busemann equation. Analogously, (2.1) is hyperbolic outside the unit sphere and elliptic inside the unit sphere. Introduce new unknown variables.

$$U_0 = (a+1)\phi - x_i\phi_{.i}, \quad U_i = \phi_{.i}, \quad i = 1, ..., n \qquad (2.2)$$

Differentiation of the first expression in (2.2) with respect to x_i (i=1,...,n) yields

$$\frac{\partial U_0}{\partial x_i} + x_j \frac{\partial U_i}{\partial x_j} - aU_i = 0 \qquad (2.3)$$

Combining (2.3) with the original equation (2.1) we can arrive at the following symmetric system

$$\alpha^i \frac{\partial U}{\partial x_i} + aU = F \qquad (2.4)$$

where

$$\alpha^1 = -\begin{bmatrix} 0 & 1 & 0 & ... & 0 \\ 1 & 0 & 0 & ... & 0 \\ 0 & 0 & 0 & ... & 0 \\ . & . & . & ... & . \\ 0 & 0 & 0 & ... & 0 \end{bmatrix} - x_1 I, \quad ...$$

$$\alpha^n = -\begin{bmatrix} 0 & 0 & 0 & ... & 1 \\ 0 & 0 & 0 & ... & 0 \\ 0 & 0 & 0 & ... & 0 \\ . & . & . & ... & . \\ 1 & 0 & . & ... & 0 \end{bmatrix} - x_n I \qquad (2.5)$$

and

$$U = (U_0, U_1, ..., U_n)^t, \quad F = (-f, 0, ..., 0)^t$$

where I is the unit matrix of order (n+1).

In what follows we shall illustrate the equivalence of (2.1) and (2.4) under some mild conditions. Suppose that U is a C^2 solution of (2.4). From (2.3) it follows that

$$x_i \frac{\partial}{\partial x_i}\left(\frac{\partial U_j}{\partial x_k} - \frac{\partial U_k}{\partial x_j}\right) + (1-a)\left(\frac{\partial U_j}{\partial x_k} - \frac{\partial U_k}{\partial x_j}\right) = 0 \; in \; \Omega, j, k = 1, ...n \qquad (2.6)$$

It is easy to see

$$\left(\frac{\partial U_j}{\partial x_k} - \frac{\partial U_k}{\partial x_j}\right) = 0, \quad in \quad \Omega$$

if $\alpha < -\frac{n}{2}$ and Ω contains the unit ball.

or

$$\left(\frac{\partial U_j}{\partial x_k} - \frac{\partial U_k}{\partial x_j}\right) = 0 \text{ on some part of } \partial\Omega \text{ emanating from which}$$

all the characteristic curves of (2.6) cover $\overline{\Omega}$ eventually.   (2.7)

Suppose the domain Ω is simply connected. Then we can find a C^2 function ϕ such that $U_i = \phi_{\cdot i}$ in Ω, i=1,...,n. From (2.3) it follows also that

$$U_0 = (a+1)\phi - x_i\phi_{\cdot i} + \text{constant, on } \Omega$$

Translating a suitable constant from ϕ and using the first equation of (2.4) we can derive ϕ satisfying (2.1). Hence the equation (2.1) is equivalent to the symmetric system (2.4) if (2.7) is valid.

Let us consider the positivity of the system (2.4). According to the definition of [FK], the criterion matrix of (2.4) is

$$C = aI - \frac{1}{2}\frac{\partial \alpha^i}{\partial x_i} = \left(a + \frac{n}{2}\right)I$$

Therefore (2.4) is positively (negatively) definite if $a > -\frac{n}{2}(a < -\frac{n}{2})$ and (2.4) ((2.4) multiplied by -1) is just the positive symmetric system. Moreover, differentiating the system (2.4) with respect to all variables s-times we can get an enlarger system which is satisfied by U and its all derivatives up to order s and is called the s-th derived system. It is not difficult to see that this system is also positively (negatively) definite if $a > -\frac{n}{2} + s(a < -\frac{n}{2})$.

Using the theory of positive symmetric system [FK] we can find many admissible boundary value problems for (2.4) from a quadratic form of U on the boundary. Denote the unit outward normal of $\partial\Omega$ by $(n_1,...,n_n)$ and $n_0 = x_i n_i$, $\beta = \alpha^i n_i$. Consider the quadratic form of U

$$U \cdot \beta U = -n_0\left\{\Sigma\left(U_i + \frac{n_i}{n_0}U_0\right)^2 + \left(1 - \frac{1}{n_0^2}\right)U_0^2\right\} \quad if \quad n_0 \neq 0$$

$$or = \tfrac{1}{2}\{(U_0 - n_i U_i)^2 - (U_0 + n_i U_i)^2\} \quad \text{otherwise}$$

Form the above expression we have the following admissible boundary conditions.

(α)

$$U_0 = U_1 =, ..., = U_n = 0$$

if $n_0 > 0$, $n_0^2 > 1$, C positive or $n_0 < 0$, $n_0^2 > 1$, C negative.

(β)

$$U_i + \left(\frac{n_i}{n_0} - \pi_i\right)U_0 = 0, \text{ where } \Sigma \pi_i^2 < \frac{1}{n_0^2} - 1$$

if $n_0 > 0$, $n_0^2 < 1$, C positive or $n_0 < 0$, $n_0^2 < 1$, C negative.

(γ)

without any boundary condition

if $n_0 < 0$, $n_0^2 > 1$, C positive or $n_0 > 0$, $n_0^2 > 1$, C negative.

(δ)

$$U_0 = \Sigma \pi_i\left(U_i + \frac{n_i}{n_0}U_0\right) \text{ where } I > \left(\left(\frac{1}{n_0^2} - 1\right)\pi_i\pi_j\right)$$

if $n_0 < 0$, $n_0^2 < 1$, C positive or $n_0 > 0$, $n_0^2 < 1$, C negative.　(2.8)

Combining (2.2) with (2.8) we have a serious boundary value problems for (2.1). And (α) corresponds to the Cauchy data, (δ) the Dirichlet, Neumann and oblique derivative data for suitable π_i and (γ) means no any boundary condition prescribed. As a particular case, let Ω be a smooth bounded domain containing the unit ball. Moreover, it is assumed that $\partial\Omega$ is space–like, i.e., $n_0 > 1$, the tangent planes to $\partial\Omega$ do not meet the unit sphere. (2.8) tells us that two kinds of boundary value problems for (2.1)

$$(\alpha) \ \phi = \frac{\partial\phi}{\partial n} = 0 \text{ on } \partial\Omega \text{ if } a > -\frac{n}{2}$$

$$(\gamma) \text{ no condition prescribed if } a < -\frac{n}{2}$$

are well-posed since after transformation (2.2), under the present circumstance (2.7) is fulfilled. With the aid of [FK] and its development [G1], Gu obtained

Theorem 1. (1) If $a > -\frac{n}{2} + s$ and $f \in H_0^s(\Omega)$ where $s > \lfloor\frac{n}{2}\rfloor + 2$, then (2.1) admits a unique $C^2(\overline{\Omega})$ solution satisfying $\phi = \frac{\partial\phi}{\partial n}$ on $\partial\Omega$. (2) If $a < -\frac{n}{2}$ and $f \in H^s(\Omega)$ where $s > \lfloor\frac{n}{2}\rfloor + 2$, then (2.1) admits one and only one $C^2(\overline{\Omega})$ solution.

Theorem 1 shows the heavy affects of the lower order terms on the formulation of the boundary value problem. In case (1) for the existence of a solution on the closed region Ω the full set of the Cauchy data should be given. This is quite different from the case of elliptic equations for which only one boundary condition needs to be prescribed. Case (2) gives an interesting example which has only one solution in $C^2(\overline{\Omega})$ without any boundary condition prescribed (indeed, [G2] verifies this assertion valid in $H^1(\Omega)$ too). Obviously, for different domain Ω and different parameter a, (2.8) will gives some other well-posed boundary value problems for (2.1). It is not difficult to get the corresponding conclusions.

The method mentioned above is also applicable to a large class of more complicated mixed equations. Consider a linear partial differential equation

$$h^{ij}\phi_{.ij} + p^i\phi_{.i} + q\phi = f \tag{2.9}$$

where h^{ij}, p^i, f are sufficiently smooth functions on $\overline{\Omega}$.

Assume that (n-1) eigenvalues of the symmetric matrix (h^{ij}) are positive whereas another eigenvalue changes its sign in $\overline{\Omega}$. By the partion of unity it is easily seen that there is a system of sufficiently smooth functions σ^i (i=1,2,...,n) such that the matrix $A = (a^{ij})$ with

$$a^{ij} = h^{ij} + \sigma^i\sigma^j \tag{2.10}$$

is positivély definite. Hence (2.9) may be rewritten in the form

$$(a^{ij} - \sigma^i\sigma^j)\phi_{.ij} + p^i\phi_{.j} + q\phi = f \tag{2.9'}$$

The elliptic and hyperbolic regions are characterized by the inequality $a_{ij}\sigma^i\sigma^j < 1$ and $a_{ij}\sigma^i\sigma^j > 1$, respectively where (a_{ij}) is the inverse matrix of (a^{ij}). Analogously, introduce

$$u_0 = \lambda\phi - \sigma^i\phi_{.i} \text{ for some constant } \lambda \text{ to be fixed} \tag{2.11}$$

$$u_i = \phi_{.i} \tag{2.12}$$

From (2.9') (2.11) (2.12) we obtain the following symmetric system of partial diferential equations of first order

$$A^i u_i + Bu = F \tag{2.13}$$

with $u = (u_0, ..., u_n)^t$, $F = (-f, 0, ..., 0)^t$ and

$$A^i = \begin{bmatrix} \sigma^i & a^{1i} & \cdots & a^{ni} \\ a^{1i} & & & \\ \cdot & & \cdot & \sigma^i A \\ \cdot & & & \\ a^{ni} & & & \end{bmatrix} \tag{2.14}$$

$$
B = - \begin{bmatrix} \frac{q}{\lambda} & \cdots & p^i - (\lambda - \frac{q}{\lambda})\sigma^i + \sigma^i\sigma^j_{\cdot j} & \cdots \\ 0 & & & \\ \cdot & & & \\ \cdot & & -\lambda a^{ij} + a^{il}\sigma^j_{\cdot l} & \\ \cdot & & & \\ 0 & & & \end{bmatrix} \tag{2.15}
$$

Under some mild condition the criterion matrix of (2.13) is positively definite. A similar discussion on the admissible boundary value problems for (2.13) as well as on the well-posed boundary value problems for (2.9) can be found in [G4].

It is noteworthy that the method used here is continuous to work for some quasi-linear equations of mixed type (see [G5]) although for them the elliptic region and the hyperbolic region cannot be identified before the solution is found.

A generalization in $\mathbb{R}^3$ of Morawetzi's result on (1.1) using the positive symmetric system is also discussed in [D2].

3 Equations of higher order

It is well known that if a PDE is invariant under an action of a Lie group, then there exist a class of special solutions which are determined by a PDE with less independent variables. In such a way we may derive many interesting and useful mixed PDEs. Let

$$
P(\xi, \tau) = \sum_{j=0}^{m} P_{m-j}(\xi)\tau^j \tag{3.1}
$$

be a homogeneous hyperblic polynomial of degree m. Here $\xi = (\xi_1, ..., \xi_n)$, $P_{m-j}(\xi)$ are homogeneous polynomials of degree m-j with real constant coefficients. Moreover, (3.1) is hyperbolic with respect to $(\xi, \tau) = (0, 1)$, i.e., as an equation of τ

$$
P(\xi, \tau) = 0 \tag{3.2}
$$

admits real roots only. No loss of generality, we suppose that P(0,1) =1. If we want to find the solution of the form $u = t^{a+1}\phi(\frac{y}{t})$ to the equation $P(\frac{\partial}{\partial y}, \frac{\partial}{\partial x})u = t^{a+1-m}f(\frac{y}{t})$, $t > 0$, $y \in \mathbb{R}^n$, then we will obtain a partial differential equation

$$
L(a)\phi = L(x, \partial_x, a)\phi = f, \quad x \in \mathbb{R}^n \tag{3.3}
$$

for the function with n independent variables. Here we identify x with $\frac{y}{t}(t > 0)$ and

$$
L(a) = \sum_{j=0}^{m} P_{m-j}(\partial_x) \prod_{k=1}^{j} (a + 2 - k - x \cdot \partial_x) \tag{3.4}
$$

Let Λ be the characteristic cone, having the origin O as its vertex. Consider $\mathbb{R}^n$ as the plane t=1 in $\mathbb{R}^{n+1}$. Then $\Sigma = \Lambda \cap \mathbb{R}^n$, the real algebraic variety which may be disconnected, is the degenerate surfaces of (3.4), namely, the number of the real rooots of the characteristic equation for (3.4) may change when the point x moves across Σ.

Let Ω be a bounded region, containing the convex hull of Σ. Suppose that $\partial\Omega$ is a smooth surface, space-like. The boundary value problems considered are as follows.

Problem T_1: $L(a)=f$, in Ω with $\phi = ... = (\frac{\partial}{\partial n})^{m-1}\phi = 0$, on $\partial\Omega$

Problem T_2: $L(a)=f$, in Ω and without any boundary condition given.

By means of the fundamental solution of hyperbolic operators with constant coefficients [G6] proves the well posedness of Problem T_i (i=1, 2).

Suppose that $p(\xi, \tau) = 0$, as an equation of τ, has a root of multiplicity l for some $\xi \neq 0$ and has no root of multiplicity $\geq l+1$ for any $\xi \neq 0$. [G6] gives

Theorem 2. If $a > m+k - \frac{n}{2} - \frac{3}{2}$ and $f \in H_0^k(\Omega)$, then in $H_0^{k+m-l}(\Omega)$ Problem T_1 has a unique solution satisfying (3.3) in the senses of distribution. If $a < -\frac{n}{2} - \frac{3}{2} - l$ and $f \in H^k(\Omega)$ with $k \geq l$, then Problem T_2 has a unique solution in $H^{k+m-l}(\Omega)$.

An explicit formula of the solutions mentioned in Theorem 2 is also obtained via the fundamental solution of the corresponding hyperbolic operator.

Remark 1. When $P(\partial_y, \partial_t)$ is the wave operator, $L(a)$ corresponds to the generalized Busemann equation (2.1) and Σ is just the unit sphere.

Remark 2. From the embedded theorem it is easy to obtain the smooth solution.

Remark 3. If $f \in C^\infty(\overline{\Omega})$, then for a generic value of the parameter a the $C^\infty(\overline{\Omega})$ solution to equation (3.3) exists uniquely, though for sufficiently large a there are quite a lot of classical solutions. The exceptional values of the parameter a, the conditions for solvability and the degree of C^∞ solutions were determined.

Remark 4. The class of equations discussed here is quite large since there are a plenty of homogeneous hyperbolic polynomials. One can list many interesting examples of solvable boundary problems.

As a continuation of the last result, Hong, J.X. considered a class of general boundary value problems for some mixed equations of higher order. So far the characteristic theory and the classification for mixed equations of higher order is not yet clear. No doubt, they are very closely related to the theory of multiple characteristics of the general differential operators and the situation is quite complicated. Hong discussed a special case where there is a pair of real characteristics reducing to a pair of complex characteristics. consider a differential operator of order m defined in a bounded domain $\Omega \subset \mathbb{R}^n$

$$P = P_m(x, D) + P_{m-1}(x, D) + ..., \quad D = \frac{1}{\sqrt{-1}}(\partial_{x_1}, ..., \partial_{x_n}) \qquad (3.5)$$

where the subscripts in the right hand of (3.5) denote the orders of corresponding differential operators, P_m is of real coefficients. Suppose that

C_1. There is a function $\phi \in C^\infty(\overline{\Omega})$ with $d\phi \neq 0$ on $\overline{\Omega}$ such that $\partial\Omega$ is the disjointed union of two smooth hypersurfaces S_{-1} and S_1, here we use the notation $S_t = \{x \in \overline{\Omega} | \phi(x) = t\}$, $t \in [-1, 1]$.

C_2. $P_m^j(x, \tau d\phi + \eta) \cdot \phi_{\cdot j} \neq 0$ for all $\tau \in \mathbb{R}^1$ and $(x, \eta) \overline{\in} S_0 \times \{0\}$ satifying $P_m^j(x, \tau d\phi + \eta) = 0$. Here $P_m^j(x, \xi) = \partial p_m / \partial \xi_j$

C_3 On S_0, $P_m^j(x, d\phi) = 0$ (j=1,...,n), $d(p_m(x, d\phi)) \cdot d\phi / |d\phi|^2 > 0$ and the matrix $P_m^{ij}(x, \phi))$ has (n-1) positive eigenvalues, here $d(P_m(x, d\phi)) = grad(P_m(x, grad\phi))$.

Frome C_1, C_2 it is easy to see P is of real principle type outside $\{(x, d\phi) | x \in S_0\}$ and there is no bicharacteristic tangent to the level surfaces $S_t (t \neq 0)$. C_3 means that S_0 is multiple characteristic surface and there occur a pair of real characteristics reducing to a pair of complex characteristics when the point x moves through S_0. Define

$$W = Re I_1(x, d\phi) + \frac{(m-1)}{2} \frac{d(P_m(x, d\phi)) \cdot d\phi}{|d\phi|^2} \tag{3.6}$$

where $I_1(x, \xi)$=principle part of $\frac{\sqrt{-1}}{2}(P(x, \xi) - P^*(x, \xi))$. This is nothing else but the subprincipal of the operator P and hence (3.8) is also coordinate invariant on multiple characteristics S_0. In view of the hypothesis C_2, on the boundary $S_a(a = \pm 1)$ we have the decomposition of the form

$$P_m(x, \tau d\phi + \eta) = P_m(x, d\phi) P_m^+(\tau) P_m^-(\tau) P_m^0(\tau)$$

$$= P_m(x, d\phi) \prod_{k=1}^{m_e(a)} (\tau - \lambda_+^k) \prod_{k=1}^{m_e(a)} (\tau - \lambda_-^k) \prod_{k=1}^{m_h(a)} (\tau - \lambda^k) \tag{3.7}$$

Here λ_+^k, λ_-^k and λ^k are the roots with positive, negative and zero imaginary parts, respectively and $2m_e(a) + m_h(a) = m$. Let $B_{j_a}(x, d\phi)$ be the boundary differential operators of order γ_{j_a} with $B_{j_a}(x, d\phi) \neq 0$ on S_a. Consider two kind of Lopatinsky conditions.

Problem T_+ (as $W > 0$):

$$\{B_{j_{-1}}(x, \tau d\phi + \eta) mod P_m^+(\tau) \ on \ S_{-1}\}_{j_{-1}} = 1, ..., m_e(-1)$$

and

$$\{B_{j_1}(x, d\phi + \eta) mod P_m^-(\tau) P_m^0(\tau) \ on \ S_{-1}\}_{j_1} = 1, ..., m_e(1) + m_h(1) \tag{3.8}$$

are linearly independent, respectively.

Problem T_- (as $W < 0$) :

$$\{B_{j_{-1}}(x, \tau d\phi + \eta) mod P_m^+(\tau) P_m^0(\tau), \text{ on } S_{-1}\} \dot{j}_{-1} = 1, ..., m_e(-1) + m_h(-1)$$

and

$$\{B_{j_1}(x, \tau d\phi + \eta) mod P_m^-(\tau), \text{ on } S_1\} \quad \dot{j}_1 = 1, ..., m_e(1) \qquad (3.9)$$

are linearly independent, respectively.

Hong, J.X. discussed the following boundary value problems $T_\pm$:

$$P(x, D)u = f, \text{ in } \Omega, \text{ with } B_{j_a}(x, D)u = g_{j_a} \text{ on } S_a \qquad (3.10)$$

and proved that (3.10) are well-posed and $Ker(T_\pm)$ are of finite dimensions. More precisely.

Theorem 3. Let $C_1 - C_3$ be fulfilled and let

$$W < 0 (or W > (m-1) \frac{d(P_m(x, d\phi))}{|d\phi|^2} \frac{d\phi}{})$$

then Problem $T_-(T_+)$ can be solved if and only if

$$(f, v) - \sum_{j_a} < g_{j_a}, C'_{j_a} v > \partial\Omega = 0, \text{ for all } v \in Ker(T_-^*)((T_+^*))$$

where C'_{j_a} are some normal systems of boundary differential operators and $T_-^*(T_+^*)$ is the adjoint problem of $T_-(T_+)$. Furthermore, the solution u to T_+ is in $H^{m-1+k}(\Omega)$ if

$$\{f, g_{j_a}\} \in H^k(\Omega) \times \prod_{j_a} H^{m-1+k-\gamma_{j_a}}(S_a)$$

whereas the solution u to T_- is also in $H^{m-1+k}(\Omega)$ if

$$\{f, g_{j_a}\} \in H^k(\Omega) \times \prod_{j_a} H^{m-1+k-\gamma_{j_a}}(S_a)$$

and

$$W + k \frac{d(P_m(x, d\phi)) \cdot d\phi}{|d\phi|^2} < 0.$$

The proof of Theorem 3 is very technical and its crucial point is to derive an energy inequality for Problem $T_-(T_+)$, the argument is carried out by the microlocal analysis (see [HO2]). There are many examples to show the case of $Ker(T_\pm) \neq \{0\}$. Refer to [G7] for the second order and to [G6] for the higher order.

4   The classification of mixed equations

Let us rewrite down a 2nd partial differential equation, as (2.9), in the general form

$$h^{ij}u_{\cdot ij} + p^i u_{\cdot i} + qu = f \quad in \quad \Omega \subset I\!\!R^n \tag{4.1}$$

Here h^{ij}, p^i are smooth smooth functions defined on $\overline{\Omega}$. Assume that

A_1: (n-1) eigenvalues of the matrix (h^{ij}) are positive while another eigenvalue λ_n vanishes on a hypersurfaces $\Sigma \subset \overline{\Omega}$ which split Ω into two parts Ω_e and Ω_h where $\lambda_n > 0$ and $\lambda_n < 0$, respectively.

(4.1) is called the second (or first) class mixed equation if Σ is (or not) its characteristic surface.

It is easily seen that Tricomi equation belongs to the first class and the generalized Busemann equation to the second class. The equations of such two classes have been studied in detail and one has known that as far as the formulations of th boundary value problems and the regularity of solutions be concerned, their behaviors are quite different. Generally speaking, in the local sense for the second class equations the coefficients of lower order terms play much important role but for the first class equations those hardly do role. However, recently it has been seen that in the global sense the situation is quite complicated. An interesting example shows that the well-posedness of a boundary value problem for a mixed equation of first class does depend on the lower order term essentially. Next we shall provide some results to illustrate these phenomena.

Beside the results stated in section 2, the theory for the mixed equations of second class has been further developed by Hong [Ho1]. Assume that

A_2: There exists a sufficiently smooth function ϕ defined in a neighborhood of Σ such that

$$\lambda_n \phi \geq 0 \ near \ \Sigma, \ d\phi \neq 0 \ and \ h^{ij}\phi_{\cdot i}\phi_{\cdot j} = 0 \ on \ \Sigma \tag{4.2}$$

A_1 and A_2 imply that (4.1) is elliptic-hyperbolic and that Σ is the characteristic as well as the degenerate surface and hence, it is the mixed equation of second class. Also assume that

A_3: $\partial\Omega = \Gamma \cup \Gamma_s$ where Γ, Γ_s are smooth enough, Γ is noncharacteristic and Γ_s space-like and there is a function $\psi \in C^\infty(\overline{\Omega}_h)$, $d\psi \neq 0$, $0 \leq \psi \leq 1$ such that $S_t = \{x \in \overline{\Omega}_h | \psi(x) = t\} 0 < t \leq 1$ are space-like and $\Gamma_s = S_1$, $\Sigma = S_0$. Moreover, any corner point $p \in \Gamma \cup \Gamma_s$ is regular, namely, there is a neighborhood, $N(p)$, such that $N(p) \cap \Omega$ is diffeomorphic to a quarter of the ball in $I\!\!R^n$.

It is evident that A_1 and A_2 imply C_3 valid on Σ. And (3.6) is written as follows

$$W = (p^i - h^{ij}_{\cdot j})\phi_{\cdot i} + \frac{h^{ij}_{\cdot k}\phi_{\cdot i}\phi_{\cdot j}\phi_{\cdot k}}{2d|\phi|^2} \tag{4.3}$$

Let us study the following boundary value problems.
T_-: As $W < 0$,

$$(L - \lambda)u = f, \text{ in } \Omega \text{ with } u = 0, \text{ on } \Gamma \text{ and } u = \frac{\partial u}{\partial n} = 0, \text{ on } \Gamma_s. \qquad (4.4)$$

T_+: As $W > 0$,

$$(L - \lambda)u = f, \text{ on } \Omega \text{ with } u = 0, \text{ on } \Gamma. \qquad (4.5)$$

Using the method of the regularization for degenerate operator [HO1] proves

Theorem 4. Let the hypotheses $A_1 - A_3$ be fulfilled. Then the problem $T_-(T_+)$ has the Fredholm behavior and there is a constant λ_0 such that it admits a unique strong solution in $H^1(\Omega)$ for any $\lambda \geq \lambda_0$ and any $f \in L^2(\Omega)$. Moreover, the solution for T_+ is always in $H^{k+1}(\Omega)$ if $f \in H^k(\Omega)$, $k \geq 1$ and the solution for T_- is in $H^{k+1}(\Omega)$ if $f \in H^k(\Omega)$, $f = ... = (\frac{\partial}{\partial n})^{k-1} f = 0$, on Γ_s and $W + k(h_k^{ij} \phi_{\cdot i} \phi_{\cdot j} \phi_{\cdot k}/|d\phi|^2) < 0$, on Σ, (k=1, ...,).

It must be emphasized that even if $f \in C^\infty(\overline{\Omega})$, without any further assumption, the regularity of solution of the problem T_- cannot be improved, e.g., u may not belong to $H^2(\Omega)$. This shows the heavy affects of the lower order terms of the equation.

Next we turn to some results on the mixed equations of the first class. The mixed equation

$$Tu = yu_{xx} + u_{yy} + au_x + bu_y + cu = f, \quad x \in \Omega \qquad (4.6)$$

is also called Tricomi equation and has been studied for long period by many authors. There have been many results on the existence or uniqueness of solutions, for instance, see [S]. In these results, certain restrictions on the geometry of the domain and on its coefficients of lower order terms of the equation were made. Sun, L.X., with less assumption, presented some results about the existence of classical solutions to Tricomi as well the generalized Tricomi boundary value problems for (4.6).

Let $\partial\Omega$ consist of three curves: the characteristic curve $\gamma_+ = BC$ emitting from point B(1,0), $\gamma_0 = AB$ smooth Jordan curve located in the upper plane with endpoints B and A(0, 0) and the remaindering curve $\gamma_- = AC$. Consider

$$(T - \lambda)u = f \text{ in } \Omega \text{ and } u = 0 \text{ on } AB \cup AC \qquad (4.7)$$

Denote the unit outward normal of $\partial\Omega$ by (n_1, n_2). Suppose that

B_1

γ_- is the characteristic and

$n_2 > 0$, $n_1 \neq 0$ at the points A and B for the curve γ_0

B_2

γ_- is noncharacteristic, $\gamma_- \cup \gamma_0$ smooth and

$n_1 < 0$, $n_2 < 0$ at the point A and $1 > 2\frac{n_2}{n_1}a(0,0)$,

$n_2 > 0$, $n_1 > 0$ at the point B.

B_1 corresponds to the Tricomi problem and B_2 the generalized Tricomi problem. By the Schwarz alternating procedure [SL1-2] gives,

Theorem 5. If $B_1(B_2)$ is fulfilled, then the problem (4.7) has the Fredholm behavior and there is a constant λ_0 such that for any given $\lambda \geq \lambda_0$ and any $f \in L^2(\Omega)$ the problem (4.7) admits a unique strong solution $u \in H^1(\Omega)$ and moreover, the solution belongs to $H^{k+1}(\Omega)$ provided that $f \in \tilde{H}^k(\Omega)(H^k(\Omega))$ for some integer $k \geq 1$.

Here $\tilde{H}^k(\Omega)$ is the space of the completion of smooth functions in $C^\infty(\overline{\Omega})$, vanishing near γ_- with the norm of $H^k(\Omega)$.

Remark 1. B_1 and B_2 are invariant under the transformation of coordinates preserving the principal part of (4.6).

Remark 2. If a(0,0) =0, there is no restriction on the slop at the point A of the curve $\gamma_- \cup \gamma_0$.

From Theorem 5 it seems that the regularity of solutions and the formulation of the problems are almost independent of the coefficients of the lower order terms, particularly for Tricomi problem. This gives a striking contrast between mixed equations of the first class and the second class. It is a pity that most obtained results on the equations of the first class are only for the two dimensional case.

The following example we shall introduce is a mixed equation for amplifying spiral wave which shows the complicated situation in the global sense. The behaviors of both two classes of mixed equations may be interwined. Along the line to derive the mixed differential operator (3.4), we seek the solution of the wave equation

$$\Phi_{xx} + \Phi_{yy} - \Phi_{tt} = 0 \qquad (4.8)$$

in such a form

$$\Phi = t^{a+1}\phi(\xi,\eta), \quad where$$

$$\xi = \frac{x\cos\omega\tau - y\sin\omega\tau}{t}, \quad \eta = \frac{x\sin\omega\tau + y\cos\omega\tau}{t} \quad \text{with } \tau = lnt \qquad (4.9)$$

These solutions behave as a power function of t of order $a+1$ along the spiral $\xi = const$, $\eta = const$ and so are called the amplifying spiral wave. Inserting (4.9) into (4.8) gives

$$L\phi = [1 - (\xi + \omega\eta)^2]\phi_{\xi\xi} - 2(\xi + \omega\eta)(\eta - \omega\xi)\phi_{\xi\eta} + [1 - (\eta - \omega\xi)^2]\phi_{\eta\eta}$$

$$+(2a + \omega^2)(\xi\phi_\xi + \eta\phi_\eta) + (2a + 1)(\eta\phi_\xi - \xi\phi_\eta) - a(a+1)\phi = f \qquad (4.10)$$

As $\omega = 0$, $a = -1$, (4.10) is just the Busemann equation. In the sequel, we always assume $\omega \neq 0$. It is easily seen that (4.10) is elliptic (or hyperbolic) if $\xi^2 + \eta^2 < 1/(1 + \omega^2)$ (or $> 1/(1 + \omega^2)$) and $\Sigma = \{\xi^2 + \eta^2 = 1/(1 + \omega^2)\}$ is the degenerate curve. Moreover, Σ is noncharacteristic and so (4.10) is the mixed equation of the first class. On the other hand, $S^1 = \{\xi^2 + \eta^2 = 1\}$ is the characteristic curve as well a limit cycle of one family of characteristic curves. This implies that this family of characteristic curves could not, through S^1, enter into the unit disk and hence in the region containing S^1, (4.10) is not global hyperbolic. Using the general method mentioned in section 2 and taking $\sigma^1 = (\xi + \omega\eta)$, $\sigma^2 = \eta - \omega\xi$ and $\lambda = (a + 1)$ we can reduce (4.10) to the following system

$$Ku = A^1 u_\xi + A^2 u_\eta + Bu = F = (-f, 0, 0)^t \qquad (4.11)$$

where

$$A^1 = -(\xi + \omega\eta)I, \qquad A^2 = -(\eta - \omega\xi)I \quad and$$

$$B = \begin{bmatrix} a & 0 & 0 \\ 0 & a & \omega \\ 0 & -\omega & 0 \end{bmatrix} \qquad (4.12)$$

Analogously, define

$$C = -\frac{1}{2}[B + B^t - A^1_\xi - A^2_\eta] \qquad (4.13)$$

It is not difficult to verify

(4.11) together with its s-th derived system are positive (negative)

symmetric if $a > -1 + s(a < -1)$ where s is a natural number

Therefore, in analogy with (2.8) we can consider the boundary condition

(α)

$$u = 0(\phi = \frac{\partial\phi}{\partial n} = 0) \quad on \quad \partial\Omega$$

(γ)

without any given condition on $\partial\Omega$.

[G8] proves

Theorem 6. Let Ω be a smooth bounded domain of the boundary where $n_0 > 0$, $n_0^2 > 1$. If $a > -1 + s$ and $f \in H_0^s(\Omega)$, then (4.10) admits a unique solution $\phi \in H^{s+1}(\Omega)$ satisfying (α). If $a < -1$ and $f \in H^1(\Omega)(H^s(\Omega))$, then (4.10) admits a unique solution $\phi \in H^2(\Omega)(H^{s+1}(\Omega))$.

The conclusion of Theorem 6 looks quite similar to that of Theorem 1. But (4.10), as mentioned above, belongs to the first class and (2.1) the second class. If we take $\Omega = \{\xi^2 + \eta^2 < R_1^2\}$ with $1/(1+\omega^2) < R_1 < 1$, then on $\partial\Omega$, $n_0 > 0$, $n_0^2 < 1$. According to Theorem 1, only one boundary condition should be given on $\partial\Omega$ for small a. In this case, (4.10) shows the behavior of the mixed equation of the first class normally.

In conclusion we should say that the theory of mixed equations still far from being completed, though we already know much more facts than thirty or fifty years ago. Besides the general theory, the applications to geometry and gasdynamics should be very essential for the further development.

References

[T] F.G. Tricomi, Sulle equazioni lineari alle derivate parziali di seconde ordinee di tipo misto, Rend. Reale Accad. Lincei, Ser 14(1923), 133–247.

[B] A. Busemann, Infinitesimale kegelige Uberschallstromungen, schriften Dtsch. Akd. Lufo. 7B (1943), 105–122.

[F] F. Frankl, On the problems of Chaplygin for mixed sub and supersonic flows, Bull. del'Acd.des Sciencede l'USSR 9(1945), 121–142.

[BE] L. Bers, Results and conjectures in the mathematical theory of subsonic and transonic gas flows, Comm. Pure Appl. Math. 7(1954).

[P1] M.H. Protter, Uniqueness theorem for the Tricomi problems I -, J. Rational Mech. Anal. 2(1953), 107–114.

[P2] M.H. Protter, Uniqueness theorem for the Trocomi problems II, J. Rational Mech. Anal., 4(1955), 721–732.

[P3] M.H. Protter, New boundary value problems for the wave equation and equations of mixed type, J. Rational Mech. Anal., 3(1954), 435–446.

[P4] M.H. Protter, Bull Amer. Math. Soc., 1(1979), 534–538.

[WD] S.M. Wu and S.S. Ding, Sur l'unicite du probleme de Tricomi de l'equation de Chaplygin, Acta Math Sinica, 5(1955), 393–349.

[W] K.Y. Wang, Sur l'unicite du probleme de Tricomi de l'equation de Chaplygin, Acta Math Sinica, 5(1955), 454–461.

[D1] G.C. Dong, Uniqueness theorem for Chaplygin problems (II), Acta Math. Sinica, 6(1956), 250–262.

[D2] G.C. Dong, Boundary value problems of a mixed equation in R^3, Jour. of Zhejiang Univ., (1965), 1–6.

[SU1] H.S. Sun, Tricomi problem for deformation of mixed curvature, Scientia Sinica, ((2)1981), 149–159.

[SU2] H.S. Sun, On the Tricomi problems of infinitesimal deformation equation for surfaces with mixed curvature, Scientia Sinica ((11) 1991), 1132– 1136.

[HU1] L.K. Hua, A talk starting from the unit circle, Science publisher Beijing, China, 1977.

[HU2] L.K. Hua, A Lavrencev like mixed equation, Acta Math. Sinica, 15(1965), 873–882.

[FK] K.O. Friedrichs, Symmetric positive linear differential equations, Comm. Pure Appl. Math., 11(1958), 333–414.

[G1] C.H. Gu, Differentiable solutions of positive symmetric systems, Acta Math. Sinica, 14(1964), 503–516.

[G2] C.H. Gu, On some differential equations of mixed type in n dimensional space, Scientia Sinica, (1965), 1574–1581.

[G3] C.H. Gu, Boundary value problems for mixed equation of n independent variables, Bull. Sciences, 67(1978), 335–339.

[G4] C.H. Gu, On partial differential equations of mixed type in n independent variables, Comm. Pure Appl. Math., 34(1981), 333–345.

[G5] C.H. Gu, Boundary value problems for quasilinear positive symmetric system and applications to mixed equations, Acta Math. Sinica, 21(1978), 119–129.

[G6] C.H. Gu, On a class of mixed partial differntial equations of higher order, Chin. Ann. of Math., 3(1982), 503–514.

[G7] C.H. Gu, On the C^∞ solutions of a class of linear partial differential equations, Comm. in PDEs, 5(10)(1980), 985–997.

[G8] C.H. Gu, The mixed equations for amplifying spiral wave, Letters in Mathematical Physics, 16(1988), 69–76.

[HO1] J.X. Hong, On boundary value problems for a class of equations of mixed type with characteristic degenerate surfaces, Chin. Ann. of Math., 2(1981), 407–424.

[HO2] J.X. Hong, Boundary value problems for differential operators with characteristic degenerate surfaces, Chin. Ann. of Math., 5(1984), 277–292.

[SL1] L.X. Sun, On boundary value problems for 2nd order mixed equations, Scientia Sinica, A(7)(1987), 673–682.

[SL2] L.X. Sun, The regularity of solutions to the generalized Tricomi problems, Acta Math. Sinica, 30(1987), 419–432.

[S] M.M. Smirnov, Equations of mixed type, AMS, Providence, R.I., 1978.

[JC] X.H. Ji and D.Q. Chen, The Tricomi's problem of the nonhomogenous equations of mixed type in the real projective plane, Mixed type of equations, Teubner-Text zur Mathematik, 90.

[DC] G.C. Dong and M.Y. Chi, Influence of Tricomi's mathematical work in china, Mixed type equations, Teubner-Text zur Mathematik, 90.

[GH] C.H. Gu and J.X. Hong, Some developments of the theory of partial differential equations of mixed type, Teubner-Text zur Mathematik, 90.

[JA] Japanese Mathematical Society, Encyclopedic dictionary of Mathematics, 1977.

Free Boundary Problems†

Lishang Jiang ††

Suzhou University, Suzhou, 215006, P.R. China

The study of the free boundary problems was started working in China as early as 1960 ([14] [19]). Since then many mathematicians have been concentrating their effects to the theoretical researches on phase change problems, filtration problems, etc.

1 Phase change problems

In 1989, Stefan established the first mathematical model for phase change problems. At the end of the 1940s Rubinstein proved the existence and uniqueness of its local solution. Thereafter many mathematicians began to concentrate their attention on how to obtaining sharp estimates so as to establish the global solvability of Stefan problem.

Following Friedman, Douglas, and Kyner, in 1962 Jiang [15] considered the following one-phase Stefan problem of nonlinear parabolic equations:

$$\frac{\partial u}{\partial t} - \frac{\partial}{\partial x}[k(x,t,u)\frac{\partial u}{\partial x}] = 0, \ 0 < x < s(t), \ 0 < t < T, \tag{1.1}$$
$$-k(0,t,u(0,t))\frac{\partial u}{\partial x}(0,t) = g(t), \quad 0 < t < T, \tag{1.2}$$
$$u(s(t),t) = 0, \quad 0 < t < T, (1.3) \tag{1.3}$$
$$-k(s(t),t,0)\frac{\partial u}{\partial x}(s(t),t) = \frac{ds(t)}{dt}, \quad 0 < t < T, \tag{1.4}$$
$$s(0) = 0. \tag{1.5}$$

Jiang [15] corrected a mistake in [31] concerning the estimate of derivative u_x and established the existence and uniqueness of the solution of boundary value problem (1.1)-(1.3) by an auxiliary function method, and so proved the existence of the global classical solution of the problem (1.1)-(1.5) using Schauder fixed point theorem. In addition, according to the generalized Hopf's boundary point lemma [15], he obtained the comparison theorem of solution for one phase Stefan problem (1.1)-(1.5). As its corollary, the uniqueness of solution and the continuous dependence of the solution upon the boundary and initial values were established.

Theorem 1.1 Suppose $k(x,t,u) \in C^6(\overline{I\!R}^+ \times \overline{I\!R}^+ \times I\!R)$, $k(x,t,w) \geq k_0 > 0$, $g(t) \in C^1(\overline{I\!R}^+)$, $g(t) \geq 0$, $g(t) \neq 0$ in the neighborhood of t=0, then

---

† 1991 Mathematics Subject Classification: 35R35

†† Supported by NNSF of China

C. Gu et al. (eds.), Partial Differential Equations in China, 67–79.
© 1994 Kluwer Academic Publishers.

1) The problem (1.1)-(1.5) has a unique classical solution and $u(x,t)$ and $s(t)$ depend continuously on $g(t)$.

2) Suppose that $\{u_i(x,t), s_i(t)\}$ are solutions of the problem (1.1)-(1.5) corresponding to $g = g_i$ (i=1,2), $g_1(t) \geq g_2(t)$, then

$$s_1(t) \geq s_2(t), \quad 0 \leq t \leq T,$$
$$u_1(x,t) \geq u_2(x,t), \quad 0 \leq x \leq s_2(t), \quad 0 \leq t \leq T.$$

In multidimensional case Duvant introduced the Baiocchi transformation to reduce the one-phase Stefan problem to a parabolic equation with discontinuous nonlinearity. Based upon this result, Chang [2] in 1983 applied his systematical research on fixed point theory of the multiple-valued mapping to the multidimensional one-phase Stefan problem, and obtained the existence of strong solutions, which provided a useful framework for the study of phase change problems.

In 1963, Jiang [16] considered the two-phase problem of a semi-infinite metallic rod when heated at one end:

$$u_{1t} - k_1 u_{1xx} = 0, \quad 0 < x < s(t),\ 0 < t \leq T, \tag{1.6}$$
$$u_{2t} - k_2 u_{2xx} = 0, \quad s(t) < x < l = +\infty,\ 0 < t \leq T, \tag{1.7}$$
$$-k_1 u_{1x}(0,t) = g(t), \quad 0 \leq t \leq T, \tag{1.8}$$
$$u_1(x,0) = \varphi_1(x), \quad 0 \leq t \leq b, \tag{1.9}$$
$$u_2(x,0) = \varphi_2(x), \quad b \leq x < l = +\infty, \tag{1.10}$$
$$u_1(s(t),t) = u_2(s(t),t) = 0, \quad 0 \leq t \leq T, \tag{1.11}$$
$$\frac{ds}{dt} = -k_1 u_{1x}(s(t),t) + k_2 u_{2x}(s(t),t), \quad 0 \leq t \leq T, \tag{1.12}$$
$$s(0) = b > 0, \tag{1.13}$$

where $u_2(x,t)$, $u_{2x}(x,t)$ are uniform bounded for $t \in [0,T]$ as $x \to +\infty$.

The free boundary $s(t)$ of two-phase Stefan problem in the general case is not monotone, which causes much trouble to get the uniform estimate of u_x. Sestini gave a set of conditions in [32]. He said that under these conditions $x=s(t)$ is monotone, increasing function, and so it is true for the L_∞-estimate of u_x and the existence of the global solution. Unfortunately, the proof in [32] about monotonicity of $s(t)$ is false. Jiang in [16] made use of Datzeff approximation method to give an approximate sequence $\{u_n(x,t), s_n(t)\}$ of the problem (1.6)-(1.13). On the basis of the strong maximum principle and under the Sestini conditions, Jiang proved the monotonicity of $s_n(t)$. Then he obtained

Theorem 1.2 If the functions $g(t)$, $\varphi_i(x)$ (i=1,2) satisfy

1) $g(t) \in C^1([0,T])$, $g(t) \leq 0$, $g'(t) \geq 0$, $0 \leq t \leq T$,

2) $\varphi_1(x) \in C^2([0,b])$, $\varphi_1(x) \geq 0$, $\varphi_1'(x) \leq 0$, $\varphi_1'' \geq 0$, $0 \leq x \leq b$,

3) $\varphi_2(x) \in C^2([0,\infty])$, $0 \geq \varphi_2(x) \geq -\Phi$, $\varphi_2'(x) \leq 0$, $\varphi_2''(x) \geq 0$,
   $0 \leq x \leq +\infty$, where Φ is a given positive constant,
   $\varphi_2(\infty) = \lim_{x \to +\infty} \varphi_2(x)$,

4) $\varphi_1(b) = \varphi_2(b) = 0$, $-\varphi_1'(b) + \varphi_2'(b) \geq 0$, $-'_1(0) = g(0)$,

then the probelm (1.6)-(1.13) has a unique classical solution on $\{0 \leq x < \infty, 0 \leq t \leq T\}$, and $\dot{s}(t) \geq 0$, $0 \leq t \leq T$.

In 1960, Oleinik gave a definition of the weak solution for one-dimensional multi-phase Stefan problem and proved the global existence and uniqueness of weak solution. Afterthen it is very interesting to know under which conditions is the weak solution of this problem a classical one ?

For the two-phase Stefan problem with the boundary value conditions on $\{0 \leq x \leq l, t > 0\}$:

$$u(0,t) = g_1(t) \geq 0, \tag{1.14}$$

$$u(l,t) = g_2(t) \leq 0, \tag{1.15}$$

Han [9] proved.

Theorem 1.3 If $g_i(t)$ (i=1,2) and $\varphi_i(x)$(i=1,2) are sufficiently smooth and satisfy monotonicity conditions:

$$g_1'(t) \geq 0, \quad g_2'(t) \leq 0,$$
$$\varphi_i'(x) \leq 0, \quad (i = 1,2)$$

then the weak solution of the Stefan problem (1.6),(1.7), (1.9)-(1.15) has to be a classical one.

In the general case, since the free boundary $s(t)$ is not monotone increasing, it is interesting to know in which way the L_∞-estimate of u_x can be obtained? Many mathematicians have paid a good deal of attention to his problem. It was solved perfectly in 1962. In [17] Jiang proved

Lemma 1.4 Suppose that $\{u(x,t), s(t)\}$ is the solution of the two-phase Stefan problem on $\overline{\Omega} = \{0 \leq x \leq l, \ 0 \leq t \leq T\}$, then the following estimate is true.

$$\max_{\overline{\Omega}} |u_x(x,t)| = \max\{\max_{0 \leq t \leq T} |\frac{\partial u_x(0,t)}{\partial x}|, \ \max_{0 \leq t \leq T} |\frac{\partial u_x(l,t)}{\partial x}|, \\ \max_{0 \leq x \leq s(0)} |u_x'(x,0)|, \ \max_{s(0) \leq x \leq l} |u_x'(x,0)|\}. \tag{1.16}$$

Its proof hinges on a short, but rather subtle, application of maximum principle.
Define an operator $s(t) = T(h(t)) = b + \int_0^t -k_1 u_{1x}(h\ (\bar{t}), \bar{t}) + k_2 u_{2x}(h(\bar{t}), \bar{t}) d\tau$ where $u_1(x,t)$, $u_2(x,t)$ is the solution of the initial boundary value problem (1.6)-(1.11). In order to make use of Leray Schauder fixed point theorem to prove the compactness of the operator $T(h)$, Jiang [17] established the accurate relationship between the regularities of the solution of heat equation and the boundary curve $x=h(t)$ in $\{0 \leq x \leq h(t), 0 \leq t \leq T\}$. For example, if $h(t) \in C_{[0,T]}^{\frac{1+\beta}{2}}$, then u_x is continuous to $x=h(t)$ and $\frac{\partial u(h(t),t)}{\partial x} \in C_{[0,T]}^{\frac{1+\beta}{2}}$ $(0 < \beta < 1)$.

By using Leray-Schauder fixed point theorem, Jiang proved

Theorem 1.5 If functions $g(t)$, $\varphi_i(x)$ (i=1,2) satisfy
1) $g(t) \in C^1([0,T])$, $g(t) \geq g_0 > 0$, $0 \leq t \leq T$,
2) $\varphi_1(x) \in C^2([0,b])$, $\varphi_1(x) \geq 0$, $0 \leq x \leq b$,

3) $\varphi_2(x) \in C^2([b, +\infty])$, $-\min(g_0 x, \Phi) \le \varphi_2(x) \le 0$, $b \le x < +\infty$,
  $\lim_{x \to +\infty} \varphi_2'(x) = 0$,
4) $\varphi_1(b) = \varphi_2(b)$, $g(0) = -\varphi_1(0)$,

where g_0, Φ are positives, then there exists a unique solution of Stefan problem
(1.6)-(1.13) in $\{0 \le x < \infty, \ 0 \le t \le T\}$.

Based on Lemma 1.4, Mai [26] proved the existence of global solution of two-phase Stefan problem with the first boundary conditions.

In [18] Jiang considered a two-phase Stefan problem for the general quasilinear parabolic equations:

$$a_1(x, t, u_1) u_{1xx} + b_1(x, t, u_1)(u_{1x})^2 + c_1(x, t, u_1) u_{1x} + d_1(x, t, u_1) u_1$$
$$-u_{1t} = f_1(x, t), \ 0 < x < s(t), \ 0 < t \le T, \tag{1.17}$$
$$a_2(x, t, u_2) u_{2xx} + b_2(x, t, u_2)(u_{2x})^2 + c_2(x, t, u_2) u_{2x} + d_2(x, t, u_2) u_2$$
$$-u_{2t} = f_2(x, t), \ 0 < x < s(t), \ 0 < t \le T, \tag{1.18}$$
$$u_1(0, t) = \psi_1(t), \ u_2(l, t) = \psi_2(t), \ 0 \le t \le T, \tag{1.19}$$
$$(1.9) - (1.13) \ hold, \tag{1.20}$$

where $0 \le b \le l < +\infty$.

Assuming that

1) a_i, b_i, c_i, d_i, f_i (i=1,2) are smooth functions with respect to their all variables, and $a_i \ge a > 0$ (i=1,2), $d_i \le c$ (i=1,2), $f_1 \le 0$, $f_2 \ge 0$,

2) $\psi_i \in C^1$(i=1,2), and $\psi_1 \ge 0$, $\psi_2 \le 0$,

3) $\varphi_i \in C^2$ (i=1,2), and $\varphi_1(0) = \varphi_2(l)$, $\varphi_1(b) = \varphi_2(b) = 0$,

4) $\psi_1(0) = \varphi_1(0)$, $\psi_2(0) = \varphi_2(l)$, $\varphi_1(b) = \varphi_2(b) = 0$,

Jiang [18] obtained the following

Theorem 1.6 Under the conditions 1)-4), the problem (1.15)-(1.18) has a classical solution; in addition to these conditions, further assume that a_i, b_i, c_i, d_i, f_i (i=1,2) are infinitely continuous differentiable, and $0 < b < l$, $\psi_1 > 0$, $\psi_2 < 0$, then the interface $s(t)$ is infinitely differentiable in $(0, T]$.

The shetch of the proof can be divided five steps:

Step 1. The existence of the local solution is proved, *i.e.* the Stefan problem (1.17)-(1.20) has a solution in $\{0 \le x \le l, \ 0 \le t \le \sigma\}$, where σ depends on L_∞-estimate of u_x at t=0. Hence it can be extended step by step up to $\{0 \le x \le l, \ 0 \le t < T_0\}$. General speaking, $T_0 < \infty$.

Step 2. Based on the comparison principle of the solution of the two-phase Stefan problem, the following inequality holds

$$\delta \psi_1(t) \le s(t) \le (l - \delta) \psi_2(t), \quad (0 \le t \le T_0),$$

where δ is independent of the lower bound of b and $|\psi_i(t)|$.

Step 3. For the quasilinear equations we generalize Lemma 1.4, and establish the estimate of derivative

$$\max_{\substack{0 \le x \le l \\ 0 \le t < T_0}} |\frac{\partial u_i}{\partial x}| \le C,$$

where C is independent of α.

Step 4. Repeating the above discussion on the local solutions, we claim $T_0 = \infty$.

Step 5. By making use of Stefan condition (1.12) and the accurate relationship between the regularity of the solution of linear parabolic equations and smoothness of the boundary curve again and again, we infer $s(t) \in C^\infty$.

In the field of the asymptotic behavior of the solution of two-phase Stefan problems, in 1961, Gu[7] considered the following two-phase Stefan problem for the general linear parabolic equations:

$$
\begin{aligned}
& u_{1xx} + B_1(x,t)u_{1x} + C_1(x,t)u_1 - A_1(x,t)u_{1t} = 0, \\
& \qquad -\infty < x < s(t),\ t > 0, & (1.21) \\
& u_{2xx} + B_2(x,t)u_{2x} + C_2(x,t)u_2 - A_2(x,t)u_{2t} = 0, & (1.22) \\
& \qquad s(t) < x < +\infty,\ t > 0, & (1.23) \\
& u_1(x,0) = \varphi_1(x), \quad -\infty < x < 0, & (1.24) \\
& u_2(x,0) = \varphi_2(x), \quad 0 < x < +\infty, & (1.25) \\
& u_1(s(t),t) = u_2(s(t),t) = 0, \quad t > 0, & (1.26) \\
& \dot{s}(t) = u_{1x}(s(t),t) - u_{2x}(s(t),t), \quad t > 0, & (1.27) \\
& s(0) = 0,
\end{aligned}
$$

where $\lim_{x \to (-1)^i \infty} A_i(x,t) = a_i^2$, $B_i(x,t) = o(\frac{1}{\sqrt{t}})$, i=1,2, and a_i (i=1,2) are constants.

By the comparison between the solution of the problem (1.21)-(1.27) and that of the two-phase Stefan problem which is corresponding to the heat equation $u_t - a_i^2 u_{xx} = 0$(i=1,2) with the initial conditions $u_1(x,0) = u_+ > 0$ $(0 < x < +\infty)$ and $u_2(x,0) = u_- < 0$ $(-\infty < x < 0)$, Gu [7] dealt with the asymptotic behavior of the solution of (1.19)-(1.25) under various assumptions on $b_i(x,t)$ and C_i (i=1,2). In 1980, Gu [8] extended the results of [7] to three-phase (*e.g.* ice-water-vapour) case and multi-phase case of Stefan problem.

In 1950, Landau established a mathematical model of a one-phase irreversible phase change problem which arises when a spacecraft is re-entering the Earth's atmosphere. In this model, free boundary composes of a series of monotone arcs or segments parallel to t-axis. On the free boundary, there are two cases that the temperature is either less or equal to a critical value. In the former the free boundary is the segment parallel to t-axis; while in the latter it is monotone arc. This implies an alternative condition imposed on the free boundary. The irreversible phase change problem is called Stefan-Signorini problem.

In 1984, Jiang and Friedman [6] considered the following Stefan-Signorini probelm:

$$u_t - u_{xx} = 0, \quad 0 < x < s(t), \ 0 < t < T, \tag{1.28}$$
$$\dot{s}(t) = u_x(s(t), t) - g(t) \leq 0, \quad 0 < t < T, \tag{1.29}$$
$$u(s(t), t) \leq 0, \quad 0 < t < T, \tag{1.30}$$
$$\dot{s}(t) u(s(t), t) = 0, \quad 0 < t < T, \tag{1.31}$$
$$u_x(0, t) = 0, \quad 0 < t < T, \tag{1.32}$$
$$u(x, 0) = \varphi(x), \quad 0 \leq x \leq 1, \ s(0) = 1. \tag{1.33}$$

They [6] made use of penalty method and proved the existence of the global classical solution. In addition, under certain conditions they estimated the number of turning points among the arcs and segments, *i.e.*, they proved

Theorem 1.7 1) Suppose that $\varphi \in C^2([0,1])$, $g \in C^1([0,T])$, $\varphi(x) \leq 0, 0 \leq x \leq 1$, $g(t) \geq 0, \ 0 \leq t \leq T_0 \leq +\infty$, T_0 is a constant or $+\infty$, and $\varphi'(1) - g(0) \leq 0$, $\varphi(1)[\varphi'(1) - g(0)] = 0$, then there exists a $T \in (0, T_0]$, such that the problem (1.28)-(1.33) has a classical solution, and if $T < T_0$, then $\lim_{t \to T_0 - 0} s(t) = 0$.

2) Further assume that there exists a positive integer m and mutually disjoint closed internals j_i (i=1,...,m),

$$\varphi'(x) - g(0) \geq 0, \quad \varphi'(x) - g(0) \not\equiv 0, \quad in \ \cup_{i=1}^m J_i,$$
$$\varphi'(x) - g(0) < 0, \quad in \ [0,1] \setminus \cup_{k=1}^{m+1} \hat{J}_k$$

then the classical solution of (1.28)-(1.33) is unique, and there exist at most m+1 closed intervals $\hat{J}_k \subset [0, T)$ (k=1, ..., m+1), such that

$$\dot{s}(t) \equiv 0 \ in \ \hat{J}_k, \ k = 1, ..., m+1,$$
$$\dot{s}(t) < 0, \ in [0,1] \setminus \cup_{k=1}^{m+1} \hat{J}_k$$

The alternative condition imposed on the free boundary of Stefan-Signorini problem causes much difficulty to prove the uniqueness. In 1986, from the penalty equation Jiang [20] proved the uniqueness of the classical solution of the problem (1.28)-(1.33), *i.e.*,

Theorem 1.8 The classical solution in the Theorem 1.7 is unique.

For the one-phase Stefan-Signorini problem of one-dimensional quasilinear parabolic equations, Liu [25] proved the existence and uniqueness of the global classical solution by the methods similar to those in [6] and [20].

In 1986, Jiang and Xie [24] considered a two-phase irreversible phase change problem:

$$u_t - k_1 u_{xx} = 0, \quad 0 < x < s(t), \ 0 < t < T, \tag{1.34}$$
$$u_t - k_2 u_{xx} = 0, \quad s(t) < x < 1, \ 0 < t < T, \tag{1.35}$$
$$\dot{s}(t) = k_1 u_x(s(t)-, t) - k_2 u_x(s(t)+, t) \leq 0, \quad 0 < t < T, \tag{1.36}$$
$$u(s(t)-, t) = u(s(t)+, t) \leq 0, \quad 0 < t < T, \tag{1.37}$$
$$\dot{s}(t) u(s(t)-, t) = 0, \quad 0 < t < T, \tag{1.38}$$
$$u_x(0, t) = 0, \quad u_x(1, t) = f(t), \ 0 \leq t \leq T, \tag{1.39}$$
$$u(x, 0) = \varphi(x), \quad 0 \leq x \leq 1, \ s(0) = b \in (0, 1), \tag{1.40}$$

where k_1, k_2, b are positive constants.

This problem is also called the two-phase Stefan-Signorini problem. It has similar properties to the one-phase Stefan-Signorini problem in physics, but is much more difficult than the latter. Jiang and Xie [24] obtained the existence of the global strong solution of the problem (1.34)-(1.40), but the existence of the global classical solutions did not be established until 1988. In that time Yu [28] suggested an alternative-Rothe method, and successfully got the uniformly estimate of second order derivative estimates with respect to x for the approximate solutions.

Theorem 1.9 Suppose $\varphi \in C^2([0,b]) \cap C^0([0,1])$, and $\varphi(x) \leq 0$, $0 \leq x \leq b$, $f \in C^1([0,T_0])$, $f(0) = \varphi(1)$, $k_1\varphi'(b-) - k_2\varphi'(b+) \leq 0$, $\varphi(b)[k_1\varphi'(b-) - k_2\varphi'(b+)] = 0$, where T_0 is positive constant or $+\infty$, then there exists T, $0 < T \leq T_0$, such that the problem (1.32)-(1.38) has a classical solution in $[0, T)$, and, either $T = T_0$ or $\lim_{t \to T-} s(t) = 0$.

2 Filtration problems

Consider the stationary water cone problem in oil reservoir with bottom water.

Suppose the region is cylindrical one. At the (r, z) plane, it shows polygonal region D=ABCDEFG, the oil-water interface (the stationary water cone) is $\Gamma = CE$, oil region is $\Omega = ABCEFG$. Let OB=R, FG=R_w, BC=H, AG=H_w (See Fig.1).

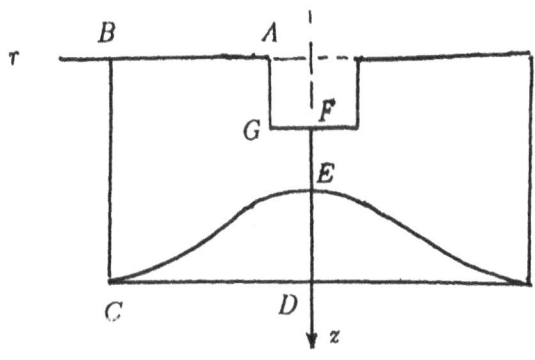

Fig. 1

Problem I. Find a triplet $\{\Omega, \rho, u\}$ such that

i) $\rho; [0, R] \to R^1$ is a strictly increasing continuous function, $\rho(R) = H$, $\rho(0) > H_w$,

ii) $\Omega = \{(r, z) \in D; r \in (0, R), 0 < z < \rho(r)\}$.

iii) $u \in C(\overline{\Omega})$, having finite energy $\int_\Omega r[(\frac{\partial u}{\partial r})^2 + (\frac{\partial u}{\partial z})^2]drdz < +\infty$, and satisfying

$$\begin{cases} \nabla \cdot (r\nabla u) = 0 \text{ in } \Omega, \\ u|_{BC} = H - z, \ u|_{AGUFG} = \alpha - z, \ u|_\Gamma = 0, \\ \int_\Omega r \cdot \nabla(u + z) \cdot \nabla\psi drdz = 0, \quad \forall(r, z) \in C_*^\infty(D), \end{cases}$$

where $\Gamma = \partial\Omega \cap D,\ 0 < \alpha < H$.

$C_*^\infty = \{\psi \in C^\infty(\overline{D});\ \psi$ vanishes in neighborhood of the closed subset $BC \cup AG \cup FG$ of $\partial D\}$.

Denote

$$\Gamma_0 = DC \cup CB \cup BA \cup AG,\quad \Gamma_N = DF \cup FG,$$
$$V^0(D) = L^2(D;r) = \{v;\ \sqrt{r}v \in L^2(D)\}$$
$$V^1(D) = \{v;\ D^\alpha v \in L^2(D;r),\ \forall|\alpha| \le 1\},$$
$$V^2(D) = \{v;\ v \in L^2(D;r),$$
$$\|u\|_{V^2}^2 = \sum_{|\alpha|=0}^2 \int_D |\partial^\alpha D|^2 r\,dr\,dz + \int_D (\tfrac{1}{r}\tfrac{\partial u}{\partial r})^2 r\,dr\,dz < +\infty\}$$

Define Baiocchi transformation

$$w(r,z) = \int_z^h [\hat\psi(r,t) - t]dt,\quad \forall(r,z) \in D,$$

where

$$\psi(r,z) = \begin{cases} \psi(r,z), & (r,z) \in \Omega, \\ z, & (r,z) \in D|_\Omega \end{cases}$$

then w satisfies

Problem II. Find a constant q and $w(r,z) \in V^2(D) \cap C^1(\overline{D})$ satifying

$$\begin{cases} \nabla \cdot (r\nabla w) = H(W) \equiv \begin{cases} 1, & w > 0 \\ 0, & w \le 0 \end{cases} \\ w|_{\Gamma_0} = g_0, \quad \frac{\partial w}{\partial n}\big|_{\Gamma_N} = g_N \end{cases}$$

where $g_0,\ g_N$ are given functions. Applying the fixed point theorems of the set-values mapping [3], Chang and Jiang [4] proved

Theorem 2.1 There exists a unique solution of Problem II.

Furthermore, by Theorem 2.1 they obtained the following

Theorem 2.2 There exists a unique solution of Problem I.

Huang and Wang proved in [13]

Theorem 2.3 $z = \rho(r)$ is analytic in $(0, R)$, and $\rho'(0) = \rho'(R) = 0$.

Consider the steady-state porous flow well problem of partially penetrating well which has an analogous mathematical form to that of the steady-state water cone problem (see Fig.2)

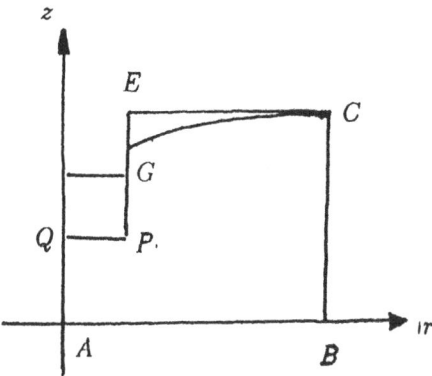

Fig. 2

1) In the case of given water level, this problem can be written as the following F.B.P.:

Problem III. Find a triplet $\{\Omega, \rho, u\}$ satisfying
   i) $\rho(r)$ is strictly increasing continuous function in $[r_0, R]$, $\rho(r_0) \geq h$, $\rho(R) = H$.
   ii) $\Omega = D_0 \cup \{(r, z) \in D; r > r_0, z < \rho(r)\}$.
   iii) $u \in V^1(\Omega) \cap C(\overline{\Omega})$, satisfying u=g on $BC \cup (EG \cap \partial\Omega) \cup GP \cup PQ$, u=0 on CF, $\int_\Omega r\nabla(u + z) \cdot \nabla v = 0$, $\forall v \in K$,
where

$$D_0 = \{(r, z) \in D; \ 0 \leq r < r_0, \ 0 < z < h_0\},$$
$$K = \{v \in V^1(\Omega); \ v = 0 \text{ on } BC \cup (EG \cap \partial\Omega) \cup GP \cup PQ\},$$
$$\text{and } g \text{ is a given function,} \tag{2.1}$$

Cryer and Zhou in [5] proved the following

Theorem 2.4 There exists a solution of Problem III.

Huang [11] proved the uniqueness:

Theorem 2.5 The solution of Problem III is unique.

2) In the case of given amount of pumped water, this stable free boundary problem of partially penetrating well can be written as the following

Problem IV. Find $\{h; \Omega, \rho, u\}$ satisfying
   i) $h_0 \leq h \leq H$,
   ii) $\rho(r)$ is a strictly increasing continuous function in $[r_0, R]$, satisfying $\rho(r_0) \geq h$, $\rho(R) = H$.
   iii) $\Omega = D_0 \cup \{(r, z) \in D; r > r_0, z < \rho(r)\}$.

iv) $u \in V^1(\Omega) \cap C(\overline{\Omega})$, satifying u=g on $BC \cup (EG \cap \partial\Omega) \cup GP \cup PQ$, u=0, on CF,

$$2\pi r_0 \int_{h_0}^{\rho(r_0)} u_r(r_0, z) dz - 2\pi \int_0^{r_0} r u_z(r, h_0) dr = Q,$$
$$\int_\Omega r\nabla(u+z) \cdot \nabla v \, dr dz = 0, \quad \forall v \in K,$$

where g is a given function.

Huang [11] proved

Theorem 2.6 There exists a unique solution of Problem IV.

Consider the non-steady compressible flow well problem, it is an evolutionary free boundary problem.

Denote the head of water by u(r,z,t), the height of free surface by $z = \varphi(r, t)$, the water level in well by c(t), c(0)=b, $0 < c(t) \le b$, $c(t) \in C^1[0, T]$, $C'(t) > -1$ (see Fig. 3).

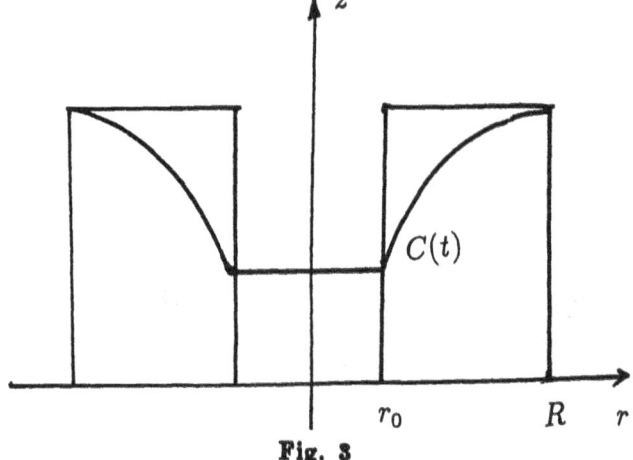

Fig. 3

The non-steady compressible flow well problem can be written as the following

Problem V. Find $\{\varphi, \Omega, u\}$ satisfying

i) $\varphi(r, t)$ is a function in $[r_0, R] \times [0, T]$,

$$\begin{cases} 0 < \varphi(r,t) \le b, & (r,t) \in [r_0, R] \times [0,T], \\ \varphi(r_0, t) \ge b, & t \in [0, T], \\ \varphi(R, t) = b, & t \in [0, T]. \end{cases}$$

ii) $\Omega = \{(r, z, t); r_0 < r < R, 0 < t < T, 0 < z < \varphi(r, t)\}$;

iii) u(r,z,t) is a smooth function on $\overline{\Omega}$, satisfying

$$\frac{1}{r}(r u_r)_r + u_{zz} - u_t = 0 \quad in \ \Omega$$

$$\begin{cases} u(r_0, z, t) = c(t), & 0 \le z \le c(t), \ 0 < t \le T \\ u(r_0, z, t) = z, & c(t) < z \le \varphi(r_0, t), \ 0 < t \le T \\ u(R, z, t) = b, & 0 \le z \le b, \ 0 < t \le T \\ u_z(r, z, t) = 0, & r_0 < r < R, \ 0 < t \le T \end{cases}$$

$$\begin{cases} u(r, z, t) = z \ on \ \Sigma \\ u_r^2 + u_z^2 - u_z = u_t, \ on \ \Sigma \end{cases}$$

where $\Sigma = \{(x, z, t); r_0 < r < R, \ 0 < t < T, \ z = \varphi(r, t)\}$, $u(r, z, 0) = b$, $r_0 \le r \le R$, $0 \le z \le b$.

Huang [11] introduced Torelli transformation:

$$w = \int_0^T \{\tilde{u}(r, z + t - \tau, \tau) - (z + t - \tau)\} d\tau$$

where $\tilde{u}$ is an extension of u, and $\tilde{u} = z$ outside of Ω, and deduced it to a variational inequality.

Problem VI. Find a function $w(x, y, t) \in K(t)$, satisfying

$w \in L^\infty(0, T; H^1(D)),$
$w|_{\Gamma_n} \in L^\infty(0, T;, H^1(\Gamma_n)),$
$w_t \in L^2(0, T; H^1(D)) \cap L^\infty(0, T; L^2(D));$
$(rw_t, v - w) + a(w, v - w) + b(\int_0^t w(\tau) d\tau, v - w) + (r, v^+ - w^+)$
$\ge - \int_{\Gamma_n} rq(v - w) dr, \quad \forall v \in K(t), \quad t \in (0, T],$

where $D = \{r_0 < r < R, 0 < z < b\}$, $Q = D \times (0, T)$, $\Gamma_n = \{r_0 < r < R, z = 0\}$, $K(t) = \{v \in H^1(D); v = g \ on \Gamma_0, \ v|_{\Gamma_n} \in H^1(\Gamma_n)\}$, g is a given function.

Huang [11] proved

Theorem 2.7 There exists a unique solution of Problem VI.

Furthermore, Wang and Huang [27] proved the regularity of w:

Theorem 2.8 $w \in H^{2+\alpha, 1+\frac{\alpha}{2}} (0 < \alpha < \frac{1}{2})$.

If the water pumped is not so much or the pumping time is not so long that we can reduce the flow well problem to a non-steady free boundary problem in the unbounded domain. Huang and Wang [12] proved the existence and uniqueness of the weak solution and that it is a limit of the solutions of the corresponding problems in the bounded domains in certain topological sense.

Chang [2] pointed out that many free boundary problems can be converted into finding the fixed points of some set-valued mappings. Investigating systematically the degree theory of a class of compact set-valued mappings, he established various fixed point theoremsm thereby setting up an effective framework for the study of the existence of the solution of free boundary problems. As one of its applications, Chang [1] proved the existence of the strong solution of two-side obstacle problem for general quasilinear elliptic equations under the condition of nature structure.

References

[1] Chang K.C., The obstacle problem and partial differential equations with discontinuous nonlinearities, Comm. Pure Appl. Math. Vol.33(1980), 117-146.

[2] Chang K.C., Free boundary problems and the set-valued mappings, J.D.E. Vol. 49(1983), No. 1, 1-28.

[3] Chang K.C. and Jiang B.J., Fixed point index of the set-valued mappings and multiplicity of solutions of elliptic equations with discontinuous nonlinearities, Acta Math. Sinica, Vol. 21(1978), 26-43.

[4] Chang K.C. and Jiang L.S., The free boundary problem of the stationary water cone, Acta Sci. Natur. Univ. Peking (1978) 1-25.

[5] Cryer C.W. and Zhou S.Z., The solution of the free boundary problem for an axisymmetric partially penetrating well, Annali Mat. Pure Appl. Ser. 4, Vol.85 (1983), 219-236.

[6] Friedman A. and Jiang L.S., A Stefan-Signorini problem, J. Diff. Equ., Vol.51 (1984), 213-231.

[7] Gu L.K., The behavior of the solution for the Stefan problem as time increase infinitely, Dokl. Akad. Nauk SSSR, Vol. 138 (1961), 263-266.

[8] Gu L.K., The asymptotic behavior of the solution of the multi-phase Stefan problem, Acta Math. Sinica, Vol.23 (1980), 203-214.

[9] Han H.T., The two-phase Stefan problem for quasilinear parabolic equations, Collection of papers on diff. equ., Peking Univ., (1963), 57-65.

[10] Huang S.Y., A free boundary problem connected with non-steady porous flow well, Acta Math. Sinica, Vol.25 (1982), 754-768.

[11] Huang S.Y., The free boundary problem for a partially penetrating well, Acta Math. Sinica, Vol.28 (1985), 71-84.

[12] Huang S.Y and Wang Y.D., A free boundary problem in an unbounded domain, Acta Math. Sinica, Vol. 26 (1983), 354-377.

[13] Huang S.Y. and Wang Y.D., The free boundary problem of the stationary water cone and its numerical approximation, Acta Sci. Natur. Peking Univ., (1983), 1-25.

[14] Jiang L.S., The free boundary problem for nonlinear parabolic equations, Acta Math. Sinica, Vol. 5(1962), 208-223.

[15] Jiang L.S., Correctness of a free boundary problem for nonlinear parabolic equations, Acta Math. Sinica, Vol.12 (1962), 369-388; translated as Chinese Math. Acta, Vol. 3 (1963), 399-418.

[16] Jiang L.S., Two phase Stefan problem (I), Acta Math. Sinica, Vol. 13 (1963), 631-646; translated as Chinese Math. Acta, Vol.4(1964), 686-702.

[17] Jiang L.S., Two-phase Stefan problem (II), Acta Math. Sinica, Vol.14 (1964), 33-49; translated as Chinese Math. Acta, Vol.5 (1964), 36-53.

[18] Jiang L.S., Existence and differentiability of the solution of the two-phase Stefan problem for quasilinear parabolic equations, Acta Math. Sinica, Vol. 15 (1965), 749-764; translated as Chinese Math. Acta, Vol. 7 (1965), 481-496.

[19] Jiang L.S., Free boundary problems in China, Numerical Treatment of Free Boundary Value Problems, J. Albrecht, L. Collatz, K. H. Hoffman, (Eds),

Birhhauser Verlag, Basel (1982), 176-186.

[20] Jiang L.S., Remarks on the Stefan-Signorini peoblem, Free Boundary Problems: Applications & Theory (III), A Bossavit, A. Damlamian, M. Fremond, (Eds), Research Notes in Math. (120) Pitman (1985), 13-19.

[21] Jiang L.S., On an elastic-plastic problem, J. Diff. Equ., Vol.51 (1984), 97-115.

[22] Jiang L.S. and Wu L.C. etc., On the existence, uniqueness of a kind of elastic-plastic problems and the convergence of the approximate solutions, Acta Math. Appl. Sinica, Vol. 4 (1981), 166-174.

[23] Jiang L.S. and Wu L.C., A class of nonlinear elliptic systems with discontinuous coefficients, Acta Math. Sinica, Vol.26 (1983), 660-668.

[24] Jiang L.S. and Xie W.Q., A two-phase Stefan-Signorini problem, Acta Sci. Natur. Peking Univ., (1986), 1-14.

[25] Liu X.Y., Stefan-Signorini problem of quasilinear parabolic equations, Acta Math. Natur. Peking Univ., (1988), 19-32.

[26] Mai M.C., Existence of the solution for two-phase Stefan problem, Collection of papers on diff. equ. Peking Univ., (1963) 45-56.

[27] Wang Y.D. and Huang S.Y., Regularity of the solution for compressible flow well problem, Acta Math. and Phys., Vol.5 (1985), 147-166.

[28] Yu W.H., Existence of solution for a two-phase Stefan-Signorini problem in one-dimensional case, J. Partial Diff. Equ., Vol.1, Ser. B, No.2, (1988), 67-83.

[29] Friedman A., One dimensional Stefan problems with nonmonotone free boundary, Trans. AMS, Vol. 133 (1968), 89-114.

[30] Friedman A., Analyticity of the free boundary of Stefan problems, Arch. Rat. Mech. Anal., Vol. 61 (1976), 97-125.

[31] Kyner W.T., An existence and uniqueness theorem for a nonlinear Stefan problem, J. Math. Mech., Vol. 8 (1959), 483-498.

[32] Sestini G., Sul problema unidimensionale non lineare di Stefan in uno strato piano indefinito, Annali Mat. Pura Appl., Vol.51 (1960), 203-224.

The Generalized Riemann Problem for Quasilinear Hyperbolic Systems of Conservation Laws [†]

Ta-tsien Li

Fudan University Shanghai 200433, China

Abstract: A general framework with some successful applications is presented in this paper for proving that the generalized Riemann problem for quasilinear hyperbolic sytems of conservation laws admits a unique local piecewise smooth solution $u = u(t, x)$ with a structure similar to a similarity solution $u = U(\frac{x}{t})$ of the corresponding Riemann problem. The corresponding global problem is also discussed in certain cases. The result shows that the similarity solution $u = U(\frac{x}{t})$ possesses a local or global nonlinear structure stability.

Key words: Quasilinear hyperbolic system of conservation laws, generalized Riemann problem, free boundary problem, nonlinear structure stability.

1 Introduction

Consider the quasilinear system of conservation laws

$$\frac{\partial u}{\partial t} + \frac{\partial f(u)}{\partial x} = 0, \tag{1.1}$$

where $u = (u_1, \cdots, u_n)^T$ is an unknown vector function of (t, x), $f : {I\!\!R}^n \to {I\!\!R}^n$ is a given smooth vector function of u. Suppose that on the domain under consideration, system (1.1) is strictly hyperbolic, i.e., the Jacobi matrix $A(u) = \nabla f(u)$ possesses n distinct real eigenvalues:

$$\lambda_1(u) < \lambda_2(u) < \cdots < \lambda_n(u). \tag{1.2}$$

Let $l_i(u) = (l_{i1}(u), \cdots, l_{in}(u))$ and $r_i(u) = (r_{i1}(u), \cdots, r_{in}(u))^T$ be the left eigenvector and the right eigenvector corresponding to $\lambda_i(u)$, respectively, for $i = 1, \cdots, n$. Without loss of generality, we may suppose that

$$l_i(u)r_j(u) \equiv \delta_{ij} \qquad (i, j = 1, \cdots, n), \tag{1.3}$$

---

† AMS subject classification: 35L65, 35L50, 35R35, 35L67.

C. Gu et al. (eds.), *Partial Differential Equations in China*, 80–103.

where δ_{ij} denotes the Kronecker's symbol.

All $\lambda_i(u)$, $l_i(u)$ and $r_i(u)$ $(i = 1, \cdots, n)$ have the same regularity as $A(u)$.

Prescribing the following piecewise constant initial condition

$$t = 0 : u = \begin{cases} \hat{u}_l, & x \le 0, \\ \hat{u}_r, & x \ge 0, \end{cases} \tag{1.4}$$

where $\hat{u}_l$ and $\hat{u}_r$ are constant vectors with

$$\hat{u}_l \ne \hat{u}_r, \tag{1.5}$$

we get Riemann problem (1.1) and (1.4), which has been studied in B.Riemann [1], R.Courant and K.O.Friedrichs [1] and P.D.Lax [1] etc. In this paper, we first give the following hypothesis:

(H1) Riemann problem (1.1) and (1.4) admits a similarity solution $u = U(\frac{x}{t})$ which consists of $n + 1$ constant states $\hat{u}_0 = \hat{u}_l, \hat{u}_1, \cdots, \hat{u}_{n-1}$ and $\hat{u}_n = \hat{u}_r$, and n waves (shocks, centered rarefaction waves or contact discontinuities) passing through the origin, the states $\hat{u}_{i-1}$ and $\hat{u}_i$ being connected by the i-th wave $(i = 1, \cdots, n)$.

Without loss of generality, we assume $n = 3$. The similarity solution $u = U(\frac{x}{t})$ can be shown, for example, in Figure 1, where S, CR and CD stand for the shock, the centered rarefaction wave and the contact discontinuity, respectively.

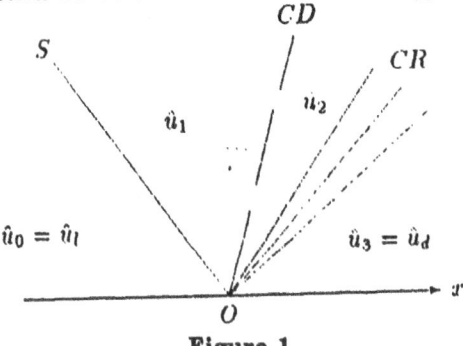

Figure 1

Remark 1.1 For some important systems of conservation laws in applications, such as the system of one-dimensional isentropic flow $(n = 2)$, the system of one-dimensional gas dynamics $(n = 3)$ etc., hypothesis (H1) is automatically satisfied and the similarity solution to Riemann problem (1.1) and (1.4) is unique (cf. R.Courant and K.O.Friedrichs [1]). For general quasilinear hyperbolic systems of conservation laws, suppose that each characteristic $\lambda_i(u)$ is either genuinely nonlinear in the sense of P.D.Lax:

$$\nabla \lambda_i(u) r_i(u) \ne 0 \tag{1.6}$$

or linearly degenerate in the sense of P.D.Lax:

$$\nabla \lambda_i(u) r_i(u) \equiv 0, \tag{1.7}$$

Riemann problem (1.1) and (1.4) admits a unique similarity solution $u = U(\frac{x}{t})$ composed of n waves with small amplitude, provided that the initial amplitude $|\hat{u}_r - \hat{u}_l|$ is suitably small (P.D.Lax [1]). In this paper we only consider the similarity solution $u = U(\frac{x}{t})$ given by (H1), no matter whether the similarity solution to Riemann problem (1.1) and (1.4) is unique or not.

We now consider the generalized Riemann problem for system (1.1) with the following piecewise smooth initial condition

$$t = 0 : u = \begin{cases} \bar{u}_l(x), & x \le 0, \\ \bar{u}_r(x), & x \ge 0, \end{cases} \qquad (1.8)$$

where $\bar{u}_l(x)$ and $\bar{u}_r(x)$ are given smooth vector functions on $x \le 0$ and on $x \ge 0$, respectively, with

$$\bar{u}_l(0) = \hat{u}_l, \qquad \bar{u}_r(0) = \hat{u}_r. \qquad (1.9)$$

Since the generalized Riemann problem (1.1) and (1.8) can be regarded as a perturbation of the corresponding Riemann problem (1.1) and (1.4), it is natural to propose the following two kinds of problems:

I. Local problem: Under which conditions does the generalized Riemann problem (1.1) and (1.8) admit a unique local piecewise smooth solution $u = u(t, x)$ which possesses a structure similar to the similarity solution $u = U(\frac{x}{t})$ of Riemann problem (1.1) and (1.4) in a neighbourhood of the origin? Here, the similar structure means that the solution $u = u(t, x)$ still consists of n waves passing through the origin and for each $i = 1, \cdots, n$, the i-th wave is of the same kind (shock, centered wave or contact discontinuity) as the i-th wave in the similarity solution $u = U(\frac{x}{t})$; moreover, at the origin the i-th wave still connects the known constant states $\hat{u}_{i-1}$ and $\hat{u}_i$ and coincides with the i-th wave in $u = U(\frac{x}{t})$. In the case $n = 3$, see Figure 2, where C denotes the centered wave.

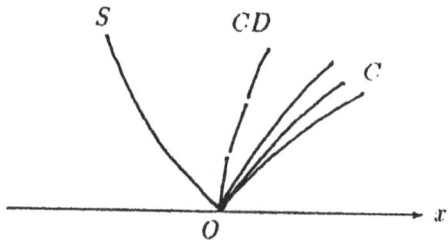

Figure 2

II. Global problem: Under which conditions does the generalized Riemann problem (1.1) and (1.8) admit a unique global piecewise smooth solution $u = u(t, x)$ on $t \ge 0$, which still consists of n waves (shocks, centered waves or contact discontinuities) passing through the origin and has a global structure similar to the similarity solution $u = U(\frac{x}{t})$ of Riemann problem (1.1) and (1.4)?

The affirmative answer of these problems shows that the similarity solution $u = U(\frac{x}{t})$ to Riemann problem (1.1) and (1.4), given by (H1), possesses a local or global nonlinear structure stability and then the piecewise smooth solution to Riemann problem (1.1) and (1.4), consisting of the preceding n waves, must be the similarity solution $u = U(\frac{x}{t})$.

For the local problem, the case that (1.1) is a single equation ($n = 1$) was discussed in A.N.Tikhonov and A.A.Samarsky [1]; the first study in the system case was done in Gu Chao-hao, Li Ta-tsien et al.[1] for the system of one-dimensional isentropic flow ($n = 2$) and in Gu Chao-hao, Li Ta-tsien and Hou Zong-yi [1-4] for the general reducible system ($n = 2$), then the system of one-dimensional gas dynamics ($n = 3$) was also discussed in Gu Chao-hao [2], Li Ta-tsien and Yu Wen-ci [5-6]. In all the situations mentioned above, the local problem has been solved for arbitrary initial amplitude $|\hat{u}_r - \hat{u}_l|$. For the general quasilinear hyperbolic system of conservation laws (1.1), suppose that each characteristic is either genuinely nonlinear or linearly degenerate in the sense of P.D.Lax, the local problem correspondng to the similarity solution $u = U(\frac{x}{t})$ with small amplitude has always an affirmative answer, provided that the initial amplitude $|\hat{u}_r - \hat{u}_l|$ is sufficiently small (see Li Ta-tsien and Yu Wen-ci [8-13]). In §2 and §3 of this paper, under hypothesis (H1) we shall present a quite general framework (with some successful applications) such that in order to solve the local probelm, we only need to check certain algebraic conditions. Finally, in §4 we shall discuss the global problem under certain additional assumptions and give the corresponding results (see Li Ta-tsien and Zhao Yan-chun [1-8]).

For the detail, the reader is referred to the following two books:

1. Li Ta-tsien and Yu Wen-ci, *Boundary Value Problems for Quasilinear Hyperbolic Systems*, Duke University Mathematics Series V, 1985 (325 pages).

2. Li Ta-tsien, *Global Classical Solutions for Quasilinear Hyperbolic Systems*, to be published in the Recherches en Mathématiques Appliquées series by Masson (315 pages).

2 Local solution to the generalized Riemann problem

For the simplicity of statement and without loss of generality, we assume $n = 3$. The general case $n \geq 2$ can be treated in a completely similar way.

In this section we first consider the situation that the similarity solution $u = U(\frac{x}{t})$ to Riemann problem (1.1) and (1.4) only consists of shocks and (or) contact discontinuities, but no centered rarefaction waves, see Figure 3. In the Figure, $\hat{u}_- = \hat{u}_1$, $\hat{u}_+ = \hat{u}_2$ and

$$O\hat{A}_i: \quad x = \hat{\sigma}_i t \quad (i = 1, 2, 3) \tag{2.1}$$

dentoe the i-th shock or the i-th contact discontinuity respectively, on which the following Rankine-Hugoniot conditions are satisfied:

$$(\hat{u}_- - \hat{u}_l)\hat{\sigma}_1 = f(\hat{u}_-) - f(\hat{u}_l), \tag{2.2}$$

$$(\hat{u}_+ - \hat{u}_-)\hat{\sigma}_2 = f(\hat{u}_+) - f(\hat{u}_-) \tag{2.3}$$

and

$$(\hat{u}_+ - \hat{u}_r)\hat{\sigma}_3 = f(\hat{u}_+) - f(\hat{u}_r). \tag{2.4}$$

Moreover, according to the entropy condition on the shock and the fact that each i-th contact discontinuity should be an i-th characteristic on both sides, noting (1.2) we have

$$\begin{cases} \lambda_1(\hat{u}_-) \leq \hat{\sigma}_1 < \lambda_2(\hat{u}_-), \\ \hat{\sigma}_1 \leq \lambda_1(\hat{u}_l), \end{cases} \tag{2.5}$$

$$\begin{cases} \lambda_1(\hat{u}_-) < \hat{\sigma}_2 \leq \lambda_2(\hat{u}_-), \\ \lambda_2(\hat{u}_+) \leq \hat{\sigma}_2 < \lambda_3(\hat{u}_+) \end{cases} \tag{2.6}$$

and

$$\begin{cases} \lambda_2(\hat{u}_+) < \hat{\sigma}_3 \leq \lambda_3(\hat{u}_+), \\ \lambda_3(\hat{u}_r) \leq \hat{\sigma}_3, \end{cases} \tag{2.7}$$

where the sign of equality corresponds to the contact discontinuity, while the sign of strict inequality corresponds to the shock. The characteristic directions marked on both sides of $O\hat{A}_i (i = 1, 2, 3)$ in Figure 3 are called the "coming" characteristic directions, where $\hat{\lambda}_i^{\pm} = \lambda_i(\hat{u}_\pm)(i = 1, 2, 3)$.

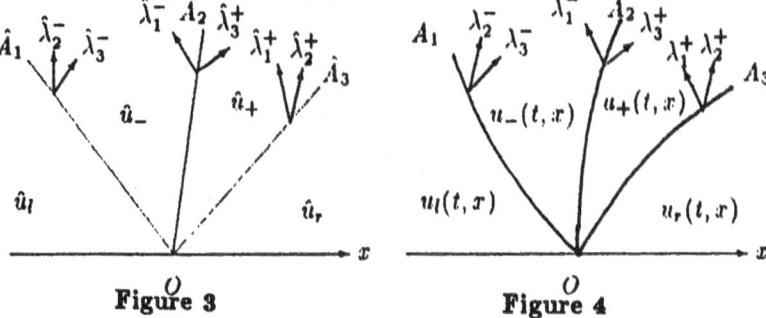

Figure 3   Figure 4

Our aim is to consider whether the generalized Riemann problem (1.4) and (1.8) admits a unique local piecewise smooth solution with a similar structure as shown in Figure 4, where for each $i = 1, 2, 3$,

$$OA_i : \quad x = x_i(t) \ (x_i(0) = 0), \tag{2.8}$$

an unknown curve passing through the origin, denotes the same kind of wave (shock or contact discontinuity) as $O\hat{A}_i$ in the solution to Riemann problem (1.1) and (1.4), and

$$x_i'(0) = \hat{\sigma}_i \quad (i = 1, 2, 3), \tag{2.9}$$

where $\hat{\sigma}_i (i = 1, 2, 3)$ are given by (2.1). The solution on both sides of $OA_i (i = 1, 2, 3)$, denoted by $u_l(t, x)$, $u_-(t, x)$, $u_+(t, x)$ and $u_r(t, x)$ respectively, satisfies system (1.1) in the classical sense and

$$u_l(0, 0) = \hat{u}_l, \ u_-(0, 0) = \hat{u}_-, \ u_+(0, 0) = \hat{u}_+, \ u_r(0, 0) = \hat{u}_r. \tag{2.10}$$

The following Rankine-Hugoniot conditions are satisfied

$$\left(u_-(t,x) - u_l(t,x)\right)\frac{dx_1(t)}{dt} = f(u_-(t,x)) - f(u_l(t,x)) \quad \text{on } x = x_1(t), \qquad (2.11)$$

$$\left(u_+(t,x) - u_-(t,x)\right)\frac{dx_2(t)}{dt} = f(u_+(t,x)) - f(u_-(t,x)) \quad \text{on } x = x_2(t) \qquad (2.12)$$

and

$$\left(u_+(t,x) - u_r(t,x)\right)\frac{dx_3(t)}{dt} = f(u_+(t,x)) - f(u_r(t,x)) \quad \text{on } x = x_3(t). \qquad (2.13)$$

Moreover, similarly to (2.5)-(2.7), we have

$$\begin{cases} \lambda_1(u_-(t,x)) \le x_1'(t) < \lambda_2(u_-(t,x)), \\ x_1'(t) \le \lambda_1(u_l(t,x)) \end{cases} \qquad \text{on } x = x_1(t), \qquad (2.14)$$

$$\begin{cases} \lambda_1(u_-(t,x)) < x_2'(t) \le \lambda_2(u_-(t,x)), \\ \lambda_2(u_+(t,x)) \le x_2'(t) < \lambda_3(u_+(t,x)) \end{cases} \qquad \text{on } x = x_2(t) \qquad (2.15)$$

and

$$\begin{cases} \lambda_2(u_+(t,x)) < x_3'(t) \le \lambda_3(u_+(t,x)), \\ \lambda_3(u_r(t,x)) \le x_3'(t) \end{cases} \qquad \text{on } x = x_3(t), \qquad (2.16)$$

where the sign of equality corresponds to the contact discontinuity, while the sign of strict inequality corresponds to the shock.

The local classical solution $u = u_l(t, x)$ on the left side of OA_1 can be obtained by solving the Cauchy problem for system (1.1) with the initial data (1.8) on $x \le 0$. In fact, when OA_1 is a 1st contact discontinuity, by (2.14) (in which we have the sign of equality) OA_1 must be a 1st characteristic passing through the origin, then the boundary of the maximal determinate domain corresponding to the initial data (1.8) on $x \le 0$; while when OA_1 is a 1st shock, by (2.14) (in which we have the sign of strict inequality) OA_1 should lie in the interior of the maximal determinate domain corresponding to the initial data (1.8) on $x \le 0$. Similarly, the local classical solution $u = u_r(t, x)$ on the right side of OA_3 can be obtained by solving the Cauchy problem for system (1.1) with the initial data (1.8) on $x \ge 0$. Obviously, we have $u_l(0,0) = \hat{u}_l$ and $u_r(0,0) = \hat{u}_r$.

Thus, in order to determine the solution to the generalized Riemann problem (1.1) and (1.8) we only need to solve the following free boundary problem for system (1.1) on the angular domain between OA_1 and OA_3: $u = u_-(t, x)$ and $u = u_+(t, x)$ satisfy system (1.1) on

$$D_-(\delta) = \{(t,x) \mid 0 \le t \le \delta, \ x_1(t) \le x \le x_2(t)\} \qquad (2.17)$$

and

$$D_+(\delta) = \{(t,x) \mid 0 \le t \le \delta, \ x_2(t) \le x \le x_3(t)\} \qquad (2.18)$$

($\delta > 0$ suitably small) respectively, and

$$u_-(0,0) = \hat{u}_-, \qquad u_+(0,0) = \hat{u}_+. \qquad (2.19)$$

Moreover, $u_-(t,x)$ and $u_+(t,x)$,   together with $u_l(t,x)$ and $u_r(t,x)$, satisfy the Rankine-Hugoniot conditions (2.11)-(2.13) and conditions (2.14)-(2.16) on $OA_i(i = 1, 2, 3)$.

Noting (2.2)-(2.4), (2.9) follows from (2.19) and (2.11)-(2.13).

Noting (2.5)-(2.7) and (2.19), by continuity the strict inequalities in (2.14)-(2.16) hold at least for a short time. Hence, we can mark the corresponding "coming" characteristic directions on both sides of $OA_i(i = 1, 2, 3)$ in Figure 4, where $\lambda_i^\pm = \lambda_i(u_\pm(t,x))(i = 1, 2, 3)$.

Let

$$u_\pm = \sum_{i=1}^{3} v_i^\pm r_i(\hat{u}_\pm). \tag{2.20}$$

Noting (1.3), we have

$$v_i^\pm = l_i(\hat{u}_\pm)u_\pm \qquad (i = 1, 2, 3). \tag{2.21}$$

We give the following hypothesis:

(H2) On $OA_i(i = 1, 2, 3)$ the Rankine-Hugoniot conditions (2.11)-(2.13) can be equivalently rewritten in an explicitly solvable form for the v variables corresponding to the "coming" characteristics. Precisely speaking, the Rankine-Hugoniot condition (2.11) on OA_1 can be rewritten as follows (noting that $u_l(t,x)$ is known): on $x = x_1(t)$

$$\frac{dx_1(t)}{dt} = F_1(t, x, u_-), \qquad x_1(0) = 0, \tag{2.22}$$

$$v_i^- = g_i^-(t, x, v_1^-) \qquad (i = 2, 3); \tag{2.23}$$

the Rankine-Hugoniot condition (2.12) on OA_2 can be rewritten as follows: on $x = x_2(t)$

$$\frac{dx_2(t)}{dt} = F_2(u_-, u_+), \qquad x_2(0) = 0, \tag{2.24}$$

$$v_1^- = g_1^-(v_2^-, v_3^-, v_1^+, v_2^+),$$
$$v_3^+ = g_3^+(v_2^-, v_3^-, v_1^+, v_2^+); \tag{2.25}$$

the Rankine-Hugoniot condition (2.13) on OA_3 can be rewritten as follows (noting that $u_r(t,x)$ is known): on $x = x_3(t)$

$$\frac{dx_3(t)}{dt} = F_3(t, x, u_+), \qquad x_3(0) = 0, \tag{2.26}$$

$$v_i^+ = g_i^+(t, x, v_3^+) \qquad (i = 1, 2). \tag{2.27}$$

Remark 2.1 In order to verify hypothesis (H2), it suffices to apply the implicit function theorem.

For the general case $n \geq 2$, suppose that OA_i $(x = x_i(t),\ x_i(0) = 0)$ is an i-th shock and the states on both sides of OA_i are denoted by $u_{i-1}(t, x)$ and $u_i(t, x)$, respectively, with $u_{i-1}(0,0) = \hat{u}_{i-1}$ and $u_i(0,0) = \hat{u}_i$. By the entropy condition, the "coming" characteristic directions on OA_i can be shown in Figure 5, where $\lambda_j^{(i-1)} = \lambda_j(u_{i-1})$ and $\lambda_j^{(i)} = \lambda_j(u_i)$ $(j = 1, \cdots, n)$. It is easy to see that if

$$\det\big(r_1(\hat{u}_{i-1}), \cdots, r_{i-1}(\hat{u}_{i-1}), \hat{u}_i - \hat{u}_{i-1}, r_{i+1}(\hat{u}_i), \cdots, r_n(\hat{u}_i)\big) \neq 0, \qquad (2.28)$$

then (H2) holds. (2.28) is actually the stability condition for an i-th shock (see A.Majda [1]).

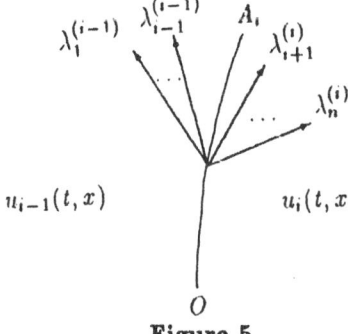

Figure 5

On the other hand, in the case that OA_i $(x = x_i(t),\ x_i(0) = 0)$ is an i-th contact discontinuity, suppose that $\lambda_i(u)$ is linearly degenerate in the sense of P.D.Lax (see (1.7)) and the Rankine-Hugoniot condition on OA_i can be equivalently rewritten as

$$w_k(u_i) = w_k(u_{i-1}) \qquad (k = 1, \cdots, i-1, i+1, \cdots, n), \qquad (2.29)$$

$$\frac{dx_i(t)}{dt} = \lambda_i(u_{i-1})(= \lambda_i(u_i)), \qquad (2.30)$$

where $w_k(u)(k = 1, \cdots, i-1, i+1, \cdots, n)$ stand for the $n-1$ independent Riemann invariants corresponding to $\lambda_i(u)$, defined by

$$\nabla w_k(u)r_i(u) = 0 \quad (k = 1, \cdots, i-1, i+1, \cdots, n). \qquad (2.31)$$

We point out that for an i-th contact discontinuity with small amplitude, the Rankine-Hugoniot condition on OA_i can be always rewritten as (2.29)-(2.30) at least in a neighbourhood of $(u_{i-1}, u_i) = (\hat{u}_{i-1}, \hat{u}_i)$, see P.D.Lax [1].

It is easily seen that if

$$\det\left(\begin{array}{cc} \nabla w_k(\hat{u}_{i-1})r_j(\hat{u}_{i-1}) & \nabla w_k(\hat{u}_i)r_j(\hat{u}_i) \\ (j = 1, \cdots, i-1) & (j = i+1, \cdots, n) \end{array}\right) \neq 0, \qquad (2.32)$$

in which $k = 1, \cdots, i-1, i+1, \cdots, n$, then (H2) holds.

Under hypothesis (H2), we form the following Jacobi matrix

$$\Theta = \frac{\partial\big(g_1^-, g_2^-, g_3^-, g_1^+, g_2^+, g_3^+\big)}{\partial\big(v_1^-, v_2^-, v_3^-, v_1^+, v_2^+, v_3^+\big)} \qquad (2.33)$$

at the point $t = x = 0$, $v^- \triangleq (v_1^-, v_2^-, v_3^-) = \hat{v}^- \triangleq (\hat{v}_1^-, \hat{v}_2^-, \hat{v}_3^-)$ and $v^+ \triangleq (v_1^+, v_2^+, v_3^+) = \hat{v}^+ \triangleq (\hat{v}_1^+, \hat{v}_2^+, \hat{v}_3^+)$, where

$$\hat{v}_i^\pm = l_i(\hat{u}_i^\pm)\hat{u}^\pm \qquad (i = 1, 2, 3). \tag{2.34}$$

Θ is called the characterizing matrix of the preceding free boundary problem on the domain $D_-(\delta) \cup D_+(\delta) \triangleq D(\delta)$. Noting (2.23), (2.25) and (2.27), the $n(n-1) \times n(n-1) = 6 \times 6$ matrix Θ possesses the following special form:

$$\Theta = \begin{pmatrix} 0 & * & * & * & * & 0 \\ * & 0 & 0 & 0 & 0 & 0 \\ * & 0 & 0 & 0 & 0 & 0 \\ 0 & 0 & 0 & 0 & 0 & * \\ 0 & 0 & 0 & 0 & 0 & * \\ 0 & * & * & * & * & 0 \end{pmatrix}. \tag{2.35}$$

Let

$$\begin{cases} \tau_1^- = \dfrac{\hat{\lambda}_1^- - \hat{\sigma}_1}{\hat{\lambda}_1^- - \hat{\sigma}_2}, & \tau_i^- = \dfrac{\hat{\lambda}_i^- - \hat{\sigma}_2}{\hat{\lambda}_i^- - \hat{\sigma}_1} \quad (i = 2, 3), \\[2mm] \tau_i^+ = \dfrac{\hat{\lambda}_i^+ - \hat{\sigma}_2}{\hat{\lambda}_i^+ - \hat{\sigma}_3} \quad (i = 1, 2), & \tau_3^+ = \dfrac{\hat{\lambda}_3^+ - \hat{\sigma}_3}{\hat{\lambda}_3^+ - \hat{\sigma}_2}, \end{cases} \tag{2.36}$$

where $\hat{\lambda}_i^\pm = \lambda_i(\hat{u}_\pm)$ $(i = 1, 2, 3)$ and $\hat{\sigma}_i$ $(i = 1, 2, 3)$ are given by (2.1). By (2.5)-(2.7) we have

$$0 \le \tau_i^\pm < 1 \qquad (i = 1, 2, 3). \tag{2.37}$$

In particular, if OA_1 is a 1st contact discontinuity, then $\tau_1^- = 0$; if OA_2 is a 2nd contact discontinuity, then $\tau_2^- = \tau_2^+ = 0$; if OA_3 is a 3rd contact discontinuity, then $\tau_3^+ = 0$.

Let

$$\Theta_k = \Theta \tau^k \qquad (k = 1, 2, \cdots), \tag{2.38}$$

where

$$\tau = \text{diag}\{\tau_1^-, \tau_2^-, \tau_3^-, \tau_1^+, \tau_2^+, \tau_3^+\}. \tag{2.39}$$

By means of the systematic result established by Li Ta-tsien and Yu Wen-ci [1-7] on free boundary problems for quaslinear hyperbolic systems, we can solve the preceding free boundary problem for system (1.1) and then the generalized Riemann problem (1.1) and (1.8).

For an $N \times N$ matrix $A = (a_{ij})$, define its minimal characterizing number

$$\|A\|_{\min} = \inf \|\gamma A \gamma^{-1}\|, \tag{2.40}$$

where the infimum is taken for all $\gamma = \text{diag}\{\gamma_1, \cdots, \gamma_N\}$ with $\gamma_i \ne 0$ $(i = 1, \cdots, N)$, and

$$\|A\| = \max_{i=1,\cdots,N} \sum_{j=1}^{N} |a_{ij}|. \tag{2.41}$$

we have

Theorem 2.1 If

$$\|\Theta_1\|_{\min} < 1, \tag{2.42}$$

then the generalized Riemann problem (1.1) and (1.8) admits a unique local piece-wise smooth solution $u = u(t, x)$ which has a structure similar to the similarity solution $u = U(\frac{x}{t})$ of the corresponding Riemann problem (1.1) and (1.4) at least in a neighbourhood of the origin $(t, x) = (0, 0)$, i.e., $u = U(\frac{x}{t})$ possesses a local nonlinear structure stability.

Remark 2.2 If $v = (v^-, v^+)$ is replaced by γv, where $\gamma = \text{diag}\{\gamma_1, \cdots, \gamma_N\}$ with $\gamma_i \neq 0$ $(i = 1, \cdots, N)$, where $N = n(n - 1) = 6$, then Θ reduces to $\gamma\Theta\gamma^{-1}$, hence Θ_1 reduces to $\gamma\Theta_1\gamma^{-1}$. Thus, Theorem 2.1 can be proved under the hypothesis

$$\|\Theta_1\| < 1 \tag{2.43}$$

instead of (2.42). (2.42) or (2.43) means that boundary conditions (2.23), (2.25) and (2.27) possess a certain dissipative property.

Condition (2.42) is only sufficient but not necessary for guaranteeing the local solvability of the generalized Riemann problem (1.1) and (1.8). In order to get a necessary and sufficient condition, successively introducing a part of tangential derivatives of the solution as new unknown functions, the original free boundary problem can be extended to a corresponding free boundary problem for a bigger hyperbolic system, then we get (cf. Li Ta-tsien and Yu Wen-ci [5-7]).

Theorem 2.2 Suppose that $f(u)$, $\bar{u}_l(x)$ and $\bar{u}_r(x)$ are sufficiently smooth. If

$$\det |I - \Theta_k| \neq 0 \qquad (k = 1, 2, \cdots), \tag{2.44}$$

then the generalized Riemann problem (1.1) and (1.8) admits a unique local piece-wise sufficiently smooth solution $u = u(t, x)$ which possesses a structure similar to the similarity solution $u = U(\frac{x}{t})$ of the corresponding Riemann problem (1.1) and (1.4) at least in a neighbourhood of the origin $(t, x) = (0, 0)$, i.e., $u = U(\frac{x}{t})$ has a local nonlinear structure stability.

Remark 2.3 Condition (2.44) is just the compatibility condition for the preceding free boundary problem on the domain $D(\delta) = D_-(\delta) \cup D_+(\delta)$. In fact, (2.44) is equivalent to the fact that all derivatives of the solution $u_\pm(t, x)$ at the origin can be uniquely determined by system (1.1) and boundary conditions (2.23), (2.25) and (2.27). Evidently, it is a necessary and sufficient condition for guaranteeing the existence and uniqueness of local piecewise sufficiently smooth solution to the generalized Riemann problem (1.1) and (1.8).

Remark 2.4 Because of (2.37), (2.44) is automatically satisfied for k large enough.

Remark 2.5 Condition (2.42) implies (2.44).

3   Local solution to the generalized Riemann problem (continued)

We now consider the situation that the similarity solution $u = U(\frac{x}{t})$ to Riemann problem (1.1) and (1.4) contains some centered rarefaction waves.

First of all, we consider the simplest case: $u = U(\frac{x}{t})$ only contains a leftmost (1st) centered rarefaction wave or (and) a rightmost (n-th) centered rarefaction wave. Without loss of generality, we still suppose $n = 3$ and consider the situation that $U(\frac{x}{t})$ only contains a leftmost centered rafefaction wave. In this case, $\lambda_1(u)$ must be genuinely nonlinear in the sense of P.D.Lax and the similarity solution $u = U(\frac{x}{t})$ can be shown in Figure 6, where $\tilde{A}_1 O \hat{A}_1$ denotes a 1st centered rarefaction wave with the origin as its center:

$$O\tilde{A}_1 : \qquad x = \lambda_1(\hat{u}_l)t, \tag{3.1}$$

$$O\hat{A}_1 : \qquad x = \lambda_1(\hat{u}_-)t \tag{3.2}$$

with

$$\lambda_1(\hat{u}_l) < \lambda_1(\hat{u}_-); \tag{3.3}$$

while for each $i = 2, 3$,

$$O\hat{A}_i : \ x = \hat{\sigma}_i t \tag{3.4}$$

stands for an i-th shock or an i-th contact discontinuity, on which the Rankine-Hugoniot condition (2.3) (resp. (2.4)) and condition (2.6) (resp. (2.7)) hold.

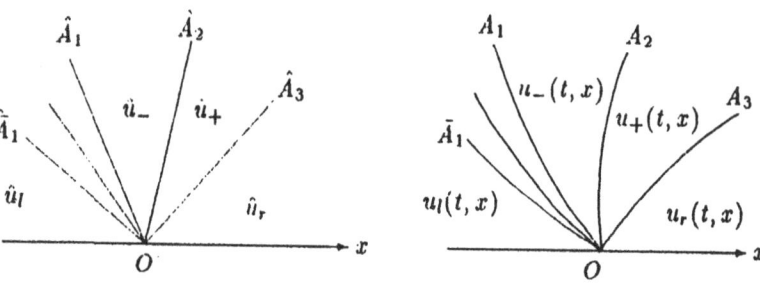

Figure 6   Figure 7

As in §2, we want to know whether the generalized Riemann problem (1.1) and (1.8) possesses a unique local piecewise smooth solution $u = u(t, x)$ with a structure similar to $u = U(\frac{x}{t})$, as shown in Figure 7, where $\bar{A}_1 O A_1$ denotes a 1st centered wave with the origin as its center (the solution $u = u_c(t, x)$ in this area is smooth for $t > 0$, but the origin is its multivalued singular point, see Li Ta-tsien and Yu Wen-ci [8,10-11]), both

$$O\bar{A}_1 : \ x = \bar{x}_1(t) \tag{3.5}$$

and

$$OA_1 : \ x = x_1(t) \tag{3.6}$$

being the 1st characteristics passing through the origin:

$$\frac{d\bar{x}_1(t)}{dt} = \lambda_1(u_c(t, \bar{x}_1(t))) = \lambda_1(u_l(t, \bar{x}_1(t))), \quad \bar{x}_1(0) = 0, \tag{3.7}$$

$$\frac{dx_1(t)}{dt} = \lambda_1(u_c(t, x_1(t))) = \lambda_1(u_-(t, x_1(t))), \quad x_1(0) = 0 \tag{3.8}$$

with

$$\bar{x}_1'(0) = \lambda_1(\hat{u}_l) \tag{3.9}$$

and

$$x_1'(0) = \lambda_1(\hat{u}_-); \tag{3.10}$$

while

$$OA_i : \ x = x_i(t) \ (x_i(0) = 0) \ (i = 2, 3) \tag{3.11}$$

with

$$x_i'(0) = \hat{\sigma}_i \quad (i = 2, 3) \tag{3.12}$$

stand for a 2nd and a 3rd shock or contact discontinuity respectively, on which the Rankine-Hugoniot conditions (2.12)-(2.13) and conditions (2.15)-(2.16) hold. Moreover, $u_l(t, x)$, $u_-(t, x)$, $u_+(t, x)$ and $u_r(t, x)$ satisfy system (1.1) in the classical sense on the corresponding domain and (2.10) holds.

As in §2, $u_l(t, x)$ and $u_r(t, x)$ can be determined by solving the Cauchy problem for system (1.1) with the initial data (1.8) on $x \le 0$ and on $x \ge 0$ respectively, and we have $u_l(0, 0) = \hat{u}_l$ and $u_r(0, 0) = \hat{u}_r$. In particular, we can determine the 1st characteristic $O\bar{A}_1(x = \bar{x}_1(t))$ as the boundary of the maximal determinate domain for the initial data on $x \le 0$, and the value of the solution on $O\bar{A}_1$:

$$u = \bar{U}_1(t) \triangleq u_l(t, \bar{x}_1(t)) \quad \text{on } x = \bar{x}_1(t) \tag{3.13}$$

with

$$\bar{U}_1(0) = \hat{u}_l. \tag{3.14}$$

By Zhou Xiao-lin [1], for any given amplitude $|\hat{u}_l - \hat{u}_-|$, using (3.13) and (3.10) we can solve a centered wave problem for system (1.1) to determine a unique 1st centered wave solution $u = u_c(t, x)$ on the domain $\bar{A}_1 O A_1$, then $O A_1(x = x_1(t))$ and the value of the solution on $O A_1$:

$$u = U_1(t) \triangleq u_c(t, x_1(t)) \quad \text{on } x = x_1(t) \tag{3.15}$$

can be uniquely defined. Thus, in order to construct the local piecewise smooth solution to the generalized Riemann problem (1.1) and (1.8), it suffices to solve a free boundary problem similar to that in §2 for system (1.1) on the domain $D(\delta) = D_-(\delta) \cup D_+(\delta)$ (still defined by (2.17)-(2.18) for $\delta > 0$ suitably small). In this free boundary problem, since $O A_1(x = x_1(t))$ is a given 1st characteristic, (2.11) and (2.14) should be replaced by (3.15) and (3.8) respectively. Furthermore, since the solution satisfies the ordinary differential equation

$$l_1(u)\frac{du}{d_1 t} = 0 \tag{3.16}$$

along each 1st characteristic, where

$$\frac{du}{d_1 t} = \frac{\partial u}{\partial t} + \lambda_1(u)\frac{\partial u}{\partial x}, \tag{3.17}$$

noting (2.20), (3.15) can be rewritten as

$$v_i^- = g_i^-(t) \quad (i = 2, 3) \qquad \text{on } x = x_1(t), \tag{3.18}$$

which will replace (2.23). Thus, the second and third rows in the characterizing matrix Θ defined by (2.33) are composed of zero elements; moreover, we have $\tau_1^- = 0$. It turns out that the matrices Θ and Θ_k have a simpler structure and Theorem 2.1 and Theorem 2.2 still hold.

We now turn our attention to the case that the similarity solution $u = U(\frac{x}{t})$ contains at least an intermediate centered rarefaction wave, say, a k-th centered rarefaction wave with $1 < k < n$, assuming that $\lambda_k(u)$ is genuinely nonlinear in the sense of P.D.Lax. Without loss of generality, we suppose $n = 3$ and $k = 2$, as shown in Figure 8, where $\hat{A}_2 O \hat{A}_2$ denotes a 2nd centered rarefaction wave with the origin as its center:

$$O\hat{A}_2: \quad x = \lambda_2(\hat{u}_-)t, \tag{3.19}$$

$$O\hat{A}_2: \quad x = \lambda_2(\hat{u}_+)t \tag{3.20}$$

with

$$\lambda_2(\hat{u}_-) < \lambda_2(\hat{u}_+); \tag{3.21}$$

while

$$OA_i: \quad x = \hat{\partial}_i t \quad (i = 1, 3) \tag{3.22}$$

stand for a 1st and a 3rd shock or contact discontinuity respectively, on which the Rankine-Hugoniot conditions (2.2) and (2.4) together with conditions (2.5) and (2.7) hold.

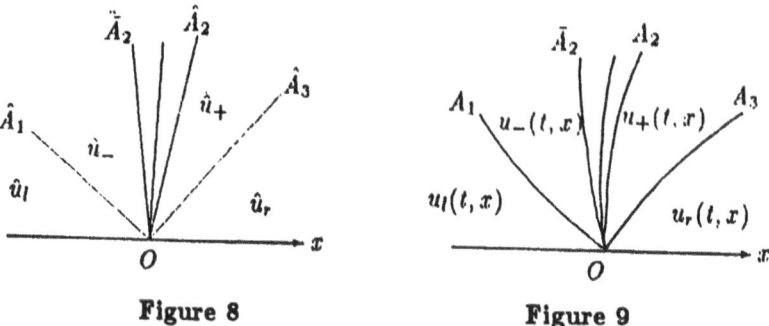

Figure 8   Figure 9

As before, we want to know whether the generalized Riemann problem (1.1) and (1.8) admits a unique local piecewise smooth solution $u = u(t, x)$ with a structure similar to $u = U(\frac{x}{t})$, as shown in Figure 9, where $\bar{A}_2 O A_2$ denotes a 2nd centered wave $u = u_c(t, x)$ with the origin as its center,

$$O\bar{A}_2: \quad x = \bar{x}_2(t) \tag{3.23}$$

and

$$OA_2 : \quad x = x_2(t) \tag{3.24}$$

being the 2nd characteristics passing through the origin:

$$\frac{d\bar{x}_2(t)}{dt} = \lambda_2(u_c(t, \bar{x}_2(t))) = \lambda_2(u_-(t, \bar{x}_2(t))), \quad \bar{x}_2(0) = 0, \tag{3.25}$$

$$\frac{dx_2(t)}{dt} = \lambda_2(u_c(t, x_2(t))) = \lambda_2(u_+(t, x_2(t))), \quad x_2(0) = 0 \tag{3.26}$$

with

$$\bar{x}_2'(0) = \lambda_2(\hat{u}_-) \tag{3.27}$$

and

$$x_2'(0) = \lambda_2(\hat{u}_+); \tag{3.28}$$

while

$$OA_i : \quad x = x_i(t) \ (x_i(0) = 0) \quad (i = 1, 3) \tag{3.29}$$

with

$$x_i'(0) = \hat{\sigma}_i \quad (i = 1, 3) \tag{3.30}$$

stand for a 1st and a 3rd shock or contact discontinuity respectively, on which the Rankine-Hugoniot conditions (2.11) and (2.13) together with conditions (2.14) and (2.16) hold. Moreover, $u_l(t, x)$, $u_-(t, x)$, $u_+(t, x)$ and $u_r(t, x)$ satisfy system (1.1) in the classical sense on the corresponding domain and (2.10) holds.

To construct the solution to the generalized Riemann problem (1.1) and (1.8), on the domain

$$D_2(\delta) = \{(t, x) \mid 0 \le t \le \delta, \ \bar{x}_2(t) \le x \le x_2(t)\} \tag{3.31}$$

($\delta > 0$ suitably small), where $x = \bar{x}_2(t)$ and $x = x_2(t)$ are two unknown 2nd characteristics passing through the origin:

$$\frac{d\bar{x}_2(t)}{dt} = \lambda_2(u(t, \bar{x}_2(t))), \quad \bar{x}_2(0) = 0, \tag{3.32}$$

$$\frac{dx_2(t)}{dt} = \lambda_2(u(t, x_2(t))), \quad x_2(0) = 0, \tag{3.33}$$

we first consider the following centered wave problem for system (1.1): the solution $u = u(t, x)$ satisfies

$$\lim_{t \to 0} u(t, \bar{x}_2(t)) = \hat{u}_-, \quad \lim_{t \to 0} u(t, x_2(t)) = \hat{u}_+ \tag{3.34}$$

and

$$v_1^+(t, x_2(t)) = \phi_1(t), \tag{3.35}$$

$$v_3^-(t, \bar{x}_2(t)) = \phi_3(t), \tag{3.36}$$

where $\phi = (\phi_1(t), \phi_3(t))$ is an arbitrarily given smooth vector function satisfying the compatibility conditions

$$\hat{v}_1^+ = \phi_1(0), \quad \hat{v}_3^- = \phi_3(0), \tag{3.37}$$

and

$$v_i^{\pm} = l_i(\hat{u}_{\pm})u, \quad \hat{v}_i^{\pm} = l_i(\hat{u}_{\pm})\hat{u}_{\pm} \quad (i = 1, 2, 3). \tag{3.38}$$

Suppose that the preceding centered wave problem admits a unique local solution $u = u(t, x)$, we may define a nonlinear operator G, called the 2nd centered wave operator, from ϕ to $\psi = (\psi_1, \psi_3)$ in the following way:

$$\psi_1(t) = v_1^-(t, \bar{x}_2(t)) \triangleq G_1[\phi], \tag{3.39}$$

$$\psi_3(t) = v_3^+(t, x_2(t)) \triangleq G_3[\phi] \tag{3.40}$$

and

$$\psi = G[\phi] = (G_1[\phi], G_3[\phi]). \tag{3.41}$$

The preceding centered wave problem can be regarded as a subproblem for solving the generalized Riemann problem (1.1) and (1.8). Precisely speaking, setting

$$\tilde{D}_2(\delta) = \{(t, x) \mid 0 \le t \le \delta, \ \tilde{\bar{x}}_2(t) \le x \le \tilde{x}_2(t)\} \tag{3.42}$$

($\delta > 0$ suitably small), we solve the following free boundary problem for system (1.1) on the domain $\{(t, x) \mid 0 \le t \le \delta, \ x_1(t) \le x \le x_3(t)\}\backslash\tilde{D}_2(\delta) \triangleq \tilde{D}(\delta)$: the piecewise smooth solution $u = \tilde{u}(t, x)$ satisfies the same conditions (2.11), (2.13) (or (2.22)-(2.23), (2.26)-(2.27)) and (2.14), (2.16) on $OA_i(x = x_i(t))$ ($i = 1, 3$) as before ($u_l(t, x)$ and $u_r(t, x)$ are still given by solving the Cauchy problem for system (1.1) with the initial data on $x \le 0$ and on $x \ge 0$ respectively), while on the free boundaries $x = \tilde{\bar{x}}_2(t)$ and $x = \tilde{x}_2(t)$, $u = \tilde{u}(t, x)$ satisfies the following conditions:

$$\lim_{t \to 0} \tilde{u}(t, \tilde{\bar{x}}_2(t)) = \hat{u}_-, \quad \lim_{t \to 0} \tilde{u}(t, \tilde{x}_2(t)) = \hat{u}_+, \tag{3.43}$$

$$\frac{d\tilde{\bar{x}}_2(t)}{dt} = \lambda_2(\tilde{u}(t, \tilde{\bar{x}}_2(t))), \quad \tilde{\bar{x}}_2(0) = 0, \tag{3.44}$$

$$\frac{d\tilde{x}_2(t)}{dt} = \lambda_2(\tilde{u}(t, \tilde{x}_2(t))), \quad \tilde{x}_2(0) = 0, \tag{3.45}$$

$$\tilde{v}_1^-(t, \tilde{\bar{x}}_2(t)) = G_1[\tilde{v}_1^+(t, \tilde{x}_2(t)), \tilde{v}_3^-(t, \tilde{\bar{x}}_2(t))], \tag{3.46}$$

$$\tilde{v}_3^+(t, \tilde{x}_2(t)) = G_3[\tilde{v}_1^+(t, \tilde{x}_2(t)), \tilde{v}_3^-(t, \tilde{\bar{x}}_2(t))], \tag{3.47}$$

where

$$\tilde{v}_i^{\pm} = l_i(\hat{u}_{\pm})\tilde{u} \quad (i = 1, 2, 3). \tag{3.48}$$

Suppose that the preceding free boundary problem for system (1.1) admits a unique local piecewise smooth solution $u = \tilde{u}(t, x)$ on the domain $\tilde{D}(\delta)$. Let

$$\phi_1(t) = \tilde{v}_1^+(t, \tilde{x}_2(t)) \tag{3.49}$$

and

$$\phi_3(t) = \tilde{v}_3^-(t, \tilde{\bar{x}}_2(t)). \tag{3.50}$$

We may solve the preceding centered wave problem (3.32)-(3.36) for system (1.1) on the domain $D_2(\delta)$. It can be proved (cf. Li Ta-tsien and Yu Wen-ci [8,11,13]) that

$$\bar{x}_2(t) \equiv \tilde{\bar{x}}_2(t), \quad x_2(t) \equiv \tilde{x}_2(t) \tag{3.51}$$

and
$$u(t, x_2(t)) \equiv \tilde{u}(t, \tilde{x}_2(t)), \qquad u(t, \bar{x}_2(t)) \equiv \tilde{u}(t, \tilde{\bar{x}}_2(t)), \tag{3.52}$$

where $\bar{x}_2(t)$, $x_2(t)$ and $u(t, x)$ are defined by the centered wave problem on $D_2(\delta)$, while $\tilde{\bar{x}}_2(t)$, $\tilde{x}_2(t)$ and $\tilde{u}(t, x)$ are defined by the free boundary problem on $\tilde{D}(\delta)$. Hence, it is easy to see that the combination of $u(t, x)$ and $\tilde{u}(t, x)$ together with $u_l(t, x)$ and $u_r(t, x)$ gives the solution to the generalized Riemann problem (1.1) and (1.8).

However, the present situation is much more complicated than the previous situation that there are only the leftmost and (or) the rightmost centered waves. In fact, in order that the preceding centered wave problem admits a unique solution $u = u(t, x)$ on $D_2(\delta)$, certain compatibility conditions should be satisfied, i.e., a two-point boundary value problem for a system of ordinary differential equations should admit a unique solution. It gives an additional requirement for the values of $\hat{u}_-$ and $\hat{u}_+$ (cf. Zhou Xiao-lin [2]). In the special case that the amplitude $|\hat{u}_+ - \hat{u}_-|$ is sufficiently small, however, the preceding compatibility conditions are automatically satisfied, and the boundary conditions of functional form (3.46)-(3.47) are close to the continuity condition on the 2nd characteristic passing through the origin:

$$u_+ = u_-. \tag{3.53}$$

Noting that the solution satisfies the ordinary differential equation

$$l_2(u) \frac{du}{d_2 t} = 0 \tag{3.54}$$

along each 2nd characteristic, where

$$\frac{du}{d_2 t} = \frac{\partial u}{\partial t} + \lambda_2(u) \frac{\partial u}{\partial x}, \tag{3.55}$$

condition (3.53) can be rewritten as

$$u_+(0, 0) = u_-(0, 0) \tag{3.56}$$

and

$$\begin{cases} v_1^- = v_1^+, \\ v_3^+ = v_3^-. \end{cases} \tag{3.57}$$

Noting (2.19), when $|\hat{u}_+ - \hat{u}_-|$ is sufficiently small, it can be proved that boundary conditions (3.46)-(3.47) are close to

$$\tilde{v}_1^- (t, \tilde{\bar{x}}_2(t)) = \tilde{v}_1^+ (t, \tilde{x}_2(t)), \tag{3.58}$$

$$\tilde{v}_3^+ (t, \tilde{\bar{x}}_2(t)) = \tilde{v}_3^- (t, \tilde{x}_2(t)), \tag{3.59}$$

hence the characterizing matrix Θ of the free boundary problem on $\tilde{D}(\delta)$ can be supposed to be of the following simple form:

$$\Theta = \begin{pmatrix} 0 & 0 & 0 & 1 & 0 & 0 \\ * & 0 & 0 & 0 & 0 & 0 \\ * & 0 & 0 & 0 & 0 & 0 \\ 0 & 0 & 0 & 0 & 0 & * \\ 0 & 0 & 0 & 0 & 0 & * \\ 0 & 0 & 1 & 0 & 0 & 0 \end{pmatrix} \tag{3.60}$$

for using Theorem 2.1 and Theorem 2.2 to get the desired conclusion (cf. Li Ta-tsien and Yu Wen-ci [13]).

To sum up, except the case that $u = U(\frac{x}{t})$ contains any intermediate centered rarefaction wave with large amplitude, in order that the generalized Riemann problem (1.1) and (1.8) admits a unique local piecewise smooth solution with a structure similar to the similarity solution $u = U(\frac{x}{t})$ of Riemann problem (1.1) and (1.4), under hypothesis (H2) it suffices to verify the algebraic condition (2.42) or (2.44) mentioned in Theorem 2.1 or Theorem 2.2.

Remark 3.1 The preceding conclusion is still valid for general strictly hyperbolic systems of conservation laws

$$\frac{\partial u}{\partial t} + \frac{\partial f(t, x, u)}{\partial x} = g(t, x, u) \tag{3.61}$$

with the corresponding Riemann problem

$$\begin{cases} \dfrac{\partial u}{\partial t} + \dfrac{\partial f(0, 0, u)}{\partial x} = 0, \\ t = 0: \ u = \begin{cases} \hat{u}_l, & x \le 0, \\ \hat{u}_r, & x \ge 0. \end{cases} \end{cases} \tag{3.62}$$

Remark 3.2 If system (1.1) or (3.61) possesses some characteristics with constant multiplicity, according to G.Boillet [1] and H.Freistüler [1], these characteristics must be linearly degenerate in the sense of P.D.Lax. Similar results on the generalized Riemann problem can be obtained in this case (cf. Li Ta-tsien [7]).

By means of the general framework mentioned above, we have successfully discussed the generalized Riemann problem for the following hyperbolic systems of conservation laws:

1. System (3.61). Suppose that (3.61) is a strictly hyperbolic system or a hyperbolic system with characteristics with constant multiplicity. Suppose furthermore that each characteristic is either genuinely nonlinear or linearly degenerate in the sense of P.D.Lax. The generalized Riemann problem possesses the desired property, provided that $|\hat{u}_r - \hat{u}_l|$ is sufficiently small, see Li Ta-tsien and Yu Wen-ci [13], Li Ta-tsien [7].

2. The system of one-dimensional isentropic flow or the general reducible hyperbolic system of conservation laws ($n = 2$) and the system of one-dimensional gas dynamics ($n = 3$) with an arbitrary initial amplitude $|\hat{u}_r - \hat{u}_l|$, see Gu Chao-hao, Li Ta-tsien et al. [1], Gu Chao-hao, Li Ta-tsien and Hou Zong-yi [1-4], Li Ta-tsien and Yu Wen-ci [5-6].

3. The system with rotation symmetry

$$\frac{\partial u}{\partial t} + \frac{\partial (f(|u|)u)}{\partial x} = 0$$

with an arbitrary initial amplitude $|\hat{u}_r - \hat{u}_l|$ (Li Ta-tsien [8]).

4. The system of the motion of elastic strings

$$\begin{cases} \dfrac{\partial u}{\partial t} - \dfrac{\partial v}{\partial x} = 0, \\ \dfrac{\partial v}{\partial t} - \dfrac{\partial}{\partial x}\Big(\dfrac{\hat{T}(r)}{r} u\Big) = 0 \end{cases} \tag{3.63}$$

with an arbitrary initial amplitude, where $u = (u_1, \cdots, u_n)$, $v = (v_1, \cdots, v_n)$ $(n \geq 2)$, $r = |u|$ and

$$\hat{T}(r) = \begin{cases} r - 1, & r \geq 1, \\ 0, & r < 1 \end{cases} \tag{3.64}$$

or

$$\hat{T}(r) = \begin{cases} T(r), & r \geq 1, \\ 0, & r < 1 \end{cases} \tag{3.65}$$

with

$$T(1) = 0, \tag{3.66}$$

$$T'(r) > \frac{T(r)}{r}, \qquad \forall\, r \geq 1 \tag{3.67}$$

and

$$T''(r) > 0, \qquad \forall\, r \geq 1, \tag{3.68}$$

see Li Ta-tsien, D.Serre and Zhang Hao [1], Li Ta-tsien and Peng Yue-jun [1].

4   Global solution to the generalized Riemann problem

Based on the result given in §2-3, we further consider whether the generalized Riemann problem (1.1) and (1.8) admits a unique global piecewise smooth solution $u = u(t, x)$ on $t \geq 0$, which still possesses a structure similar to the similarity solution $u = U(\frac{x}{t})$ of the corresponding Riemann problem (1.1) and (1.4), i.e., $u = u(t, x)$ still consists of n waves passing through the origin, just as in the local solution given in §2-3, and no new singularities occur on $t \geq 0$.

For this purpose, the key point is to establish, under suitable additional hypotheses, a uniform a priori estimate for the C^1 norm of the solution on each regular domain.

For the system of one-dimensional isentropic flow $(n = 2)$, the corresponding result can be found in Li Ta-tsien and Zhao Yan-chun [2-5].

For the generalized Riemann problem for general quasilinear strictly hyperbolic systems of conservation laws

$$\frac{\partial u}{\partial t} + \frac{\partial f(u)}{\partial x} = g(u), \tag{4.1}$$

$$t = 0: \quad u = \begin{cases} \bar{u}_l(x), & x \leq 0, \\ \bar{u}_r(x), & x \geq 0, \end{cases} \tag{4.2}$$

where $f(u)$, $g(u)$, $\bar{u}_l(x)$ and $\bar{u}_r(x)$ are suitably smooth on the domain under consideration and

$$g(0) = 0, \qquad \nabla g(0) = 0, \tag{4.3}$$

up to now we can only consider the case that the similarity solution $u = U(\frac{x}{t})$ to the corresponding Riemann problem (1.1) and (1.4) consists of n shocks. Still taking $n = 3$ as an example, the similarity solution $u = U(\frac{x}{t})$ is shown in Figure 3, where $O\hat{A}_i(x = \hat{\sigma}_i t)$ $(i = 1, 2, 3)$ are all shocks, hence the sign of strict inequality holds in (2.5)-(2.7). We want to get a unique global piecewise smooth solution $u = u(t, x)$ to the generalized Riemann problem (4.1)-(4.2) on $t \geq 0$, as shown in Figure 4, where $OA_i(x = x_i(t))$ $(i = 1, 2, 3)$ are all shocks and the sign of strict inequality holds in (2.14)-(2.16).

To this end, besides (H2) we give the following additional hypotheses on the initial data:

$$|\bar{u}_l(x) - \hat{u}_l|, \quad |\bar{u}_l'(x)| \leq \frac{\varepsilon}{1 + |x|}, \quad \forall\, x \leq 0, \tag{4.4}$$

$$|\bar{u}_r(x) - \hat{u}_r|, \quad |\bar{u}_r'(x)| \leq \frac{\varepsilon}{1 + |x|}, \quad \forall\, x \geq 0, \tag{4.5}$$

where $\varepsilon > 0$ is a small parameter. We have (cf. Li Ta-tsien and Zhao Yan-chun [6-7])

Theorem 4.1 Under the preceding hypotheses, if

$$\|\Theta_{-1}\|_{\min} < 1, \tag{4.6}$$

where

$$\Theta_{-1} = \tau^{-1}\Theta \tag{4.7}$$

and τ^{-1} stands for the inverse matrix of the diagonal matrix τ given by (2.39), then the generalized Riemann problem (4.1)-(4.2) admits a unique global piecewise smooth solution $u = u(t, x)$ on $t \geq 0$, which possesses a global structure similar to that of the similarity solution $u = U(\frac{x}{t})$ to the corresponding Riemann problem (1.1) and (1.4), hence $u = U(\frac{x}{t})$ has a global nonlinear structure stability.

Precisely speaking, $u = u(t, x)$ is composed of $n(= 3)$ shocks OA_i $(x = x_i(t)$, $x_i(0) = 0)$ $(i = 1, 2, 3)$ with

$$x_i'(0) = \hat{\sigma}_i \quad (i = 1, 2, 3) \tag{4.8}$$

$$u_l(0, 0) = \hat{u}_l, \quad u_-(0, 0) = \hat{u}_-, \quad u_+(0, 0) = \hat{u}_+, \quad u_r(0, 0) = \hat{u}_r; \tag{4.9}$$

moreover,

$$|x_i'(t) - \hat{\sigma}_i| \leq \frac{K\varepsilon}{1 + t} \quad (i = 1, 2, 3), \quad \forall\, t \geq 0 \tag{4.10}$$

and on the domain under consideration

$$|u_l(t, x) - \hat{u}_l|, \quad |\frac{\partial u_l}{\partial x}(t, x)|, \quad |\frac{\partial u_l}{\partial t}(t, x)| \leq \frac{K\varepsilon}{1 + t}, \quad \forall\, t \geq 0, \tag{4.11}$$

$$|u_-(t, x) - \hat{u}_-|, \quad |\frac{\partial u_-}{\partial x}(t, x)|, \quad |\frac{\partial u_-}{\partial t}(t, x)| \leq \frac{K\varepsilon}{1 + t}, \quad \forall\, t \geq 0, \tag{4.12}$$

$$|u_+(t, x) - \hat{u}_+|, \quad |\frac{\partial u_+}{\partial x}(t, x)|, \quad |\frac{\partial u_+}{\partial t}(t, x)| \leq \frac{K\varepsilon}{1 + t}, \quad \forall\, t \geq 0 \tag{4.13}$$

and

$$|u_r(t, x) - \hat{u}_r|, \ |\frac{\partial u_r}{\partial x}(t, x)|, \ |\frac{\partial u_r}{\partial t}(t, x)| \le \frac{K\varepsilon}{1 + t}, \quad \forall \, t \ge 0, \qquad (4.14)$$

where K is a positive constant.

Remark 4.1 Suppose that all the characteristics of system (4.1) are genuinely nonlinear in the sense of P.D.Lax and

$$\eta \triangleq |\hat{u}_r - \hat{u}_l| = |\bar{u}_r(0) - \bar{u}_l(0)| > 0 \qquad (4.15)$$

is sufficiently small. By P.D.Lax [1] the Riemann problem (1.1) and (1.4) admits a unique similarity solution with small amplitude $u = U(\frac{x}{t})$ and (H2) is automatically satisfied. Moreover, we have

$$\|\Theta_{-1}\|_{\min} = O(\eta), \qquad (4.16)$$

i.e., (4.6) holds. Then by Theorem 4.1, $u = U(\frac{x}{t})$ composed of n shocks possesses a global nonlinear structure stability.

Remark 4.2 For a global asymptotic expansion of the global piecewise smooth solution given by Theorem 4.1 and Remark 4.1, see Ph. Le Floch and Li Ta-tsien [1-2].

References

G.Boillat
[1] Chocs caractéristiques, C.R.Acad. Sci. Paris, Sér. A, 274(1972), 1018-1021.

R.Courant and K.O.Friedrichs
[1] *Supersonic flow and shock waves*, New York, Interscience, 1948.

H.Freistühler
[1] Linear degeneracy and shock waves, Math. Z., 207(1991), 583-596.

Gu Chao-hao
[1] A method for solving the supersonic flow past a curved wedge (in Chinese), Fudan Journal (Natural Science), 7(1962), 11-14.
[2] A boundary value problem of hyperbolic systems and its applications (in Chinese), Acta Mathematica Sinica, 13(1963), 32-48.

Gu Chao-hao, Li Ta-tsien et al.
[1] The Cauchy problem of typical hyperbolic systems with discontinuous initial values (in Chinese), Collections of Mathematical Papers of Fudan University, 1960, 1-16.

[2] Supersonic flow past a curved wedge (in Chinese), Collection of Mathematical Papers of Fudan University, 1960, 17-28.

Gu Chao-hao, Li Ta-tsien and Hou Zong-yi

[1] The Cauchy problem of hyperbolic systems with discontinuous initial values (in Chinese), Collections of Scientific and Technological Papers, Shanghai, 1960, I.55-I.65.

[2] The Cauchy problem for quasilinear hyperbolic systems with discontinuous initial data (I) (in Chinese), Acta Mathematica Sinica, 11(1961), 314-323.

[3] The Cauchy problem for quasilinear hyperbolic systems with discontinuous initial data (II) (in Chinese), Acta Mathematica Sinica, 11(1961), 324-327.

[4] The Cauchy problem for quasilinear hyperbolic systems with discontinuous initial data (III) (in Chinese), Acta Mathematica Sinica, 12(1962), 132-143.

Hsiao Ling and Chang Tong

[1] Perturbations of the Riemann problem in gas dynamics, J. Math. Anal. Appl., 79(1981), 436-460.

P.D.Lax

[1] Hyperbolic systems of conservation laws II, Comm. Pure Appl. Math., 10(1957), 537-556.

Ph. Le Floch and Li Ta-tsien

[1] Un développement asymptotique défini globalement en temps pour la solution du problème de Riemann généralisé, C. R. Acad. Sci. Paris, 309, Série 1 (1989), 807-810.

[2] A global asymptotic expansion for the solution to the generalized Riemann problem, Asymptotic Analysis, 3(1991), 321-340.

Li Ta-tsien

[1] Une remarque sur un problème à frontière libre, C. R. Acad. Sci. Paris, 289, Série A(1979), 99-102.

[2] On a free boundary problem, Chin. Ann. of Math., 1(1980), 351-358.

[3] Problèmes aux limites et solutions discontinues pour les systèmes hyperboliques quasi linéaires d'ordre 1, Collège de France Seminal, Vol.2: Nonlinear partial differential equations and their applications (edited by H.Brézis and J.L.Lions), Pitman, 1982, 265-302.

[4] Some free boundary problems for quasilinear hyperbolic systems, Free Boundary Problems: Applications and Theory, Vol. IV (edited by A.Bossavit, A.Damlamian and M.Fremond), Research Notes in Mathematics 121, Pitman, 1985, 582-592.

[5] Global solutions for some free boundary problems for quasilinear hyperbolic systems and applications, Nonlinear Hyperbolic Problems (Proceedings, St. Etienne 1986, edited by C.Carasso, P.A.Raviart and D.Serre), Lecture Notes in Mathematics 1270, Springer-Verlag, 1987, 195-209.

[6] Propagation of shocks and globally defined classical discontinuous solutions for quasilinear hyperbolic systems, Free Boundary Problems: Theory and Applications, Volume II (edited by K.H.Hoffmann and J.Sprekels), Pitman Research Notes in Mathematics Series 186, Longman Scientific & Technical, 1990, 747-749.

[7] A note on the generalized Riemann problem, Acta Mathematica Scientia, 11(1991), 283-289.

[8] The generalized Riemann problem for quasilinear hyperbolic systems of conservation laws with rotation symmetry, unpublished (1990).

Li Ta-tsien and Peng Yue-jun

[1] Le problème de Riemann généralisé pour une sorte de systèmes des cables, to appear in Portugaliae Mathematica.

Li Ta-tsien, D.Serre and Zhang Hao

[1] The generalized Riemann problem for the motion of elastic strings, SIAM J. Math. Anal., 23(1992), 1189-1203.

Li Ta-tsien and Yu Wen-ci

[1] Boundary value problems and discontinuous solutions for quasilinear hyperbolic systems (in Chinese), Collections of Mathematical Papers of Fudan University, 1964, 59-94.

[2] Some existence theorems for quasilinear hyperbolic systems of partial differential equations in two independent variables I: Typical boundary value problems, Scientia Sinica 4(1964), 529-550.

[3] Some existence theorems for quasilinear hyperbolic systems of partial differential equations in two independent variables II: Typical boundary value problems of functional form and typical free boundary problems, Scientia Sinica, 4(1964), 551-562.

[4] Some existence theorems for quasilinear hyperbolic systems of partial differential equations in two independent variables III: Gerneral boundary value problems and general free boundary problems, Scientia Sinica, 7(1965), 1065-1067; Fudan Journal (Natural Science), 2-3(1965), 113-128.

[5] The local solvability of the boundary value problems for quasilinear hyperbolic systems of partial differential equations, Scientia Sinica, 11(1980), 1357-1367.

[6] Boundary value problems for the first order quasilinear hyperbolic systems and their applications, J. of Diff. Equations, 41(1981), 1-26.

[7] On the necessity of the solvable conditions of the typical boundary value problem for quasilinear hyperbolic systems, Comm. in PDE., 6(1981), 1225-1234.

[8] Problèmes d'onde centrée pour les systèmes hyperboliques quasi linéaires et applications, C. R. Acad. Sci. Paris, 299, Série I (1984), 375-378.

[9] On the discontinuity relations for quasilinear hyperbolic systems (in Chinese), Journal of Engineering Mathematics, Vol.2, 2(1985), 1-10.

[10] Centered wave solutions for quasilinear hyperbolic systems (in Chinese), Fudan Journal (Natural Science), 25(1986), 195-206.

[11] Centered wave problems for quasilinear hyperbolic systems (in Chinese), Cin. Ann. of Math., 7A(1986), 423-436.

[12]   Mixed initial-boundary value problems with characteristic initial manifold for a class of quasilinear hyperbolic systems, Fudan Journal (Natural Science), 26(1987), 1-9.

[13] The discontinuous initial value problem for quasilinear hyperbolic systems (in Chinese), Journal of Engineering Mathematics, Vol.4, 2(1987), 1-12.

Li Ta-tsien and Zhao Yan-chun

[1] Vacuum problems for the system of one dimensional isentropic flow (in Chinese), Chinese Quarterly Journal of Mathematics, 1(1986), 41-46.

[2] Globally defined classical solutions to free boundary problems with characteristic boundary for quasilinear hyperbolic systems, Chin. Ann. of Math., 9B(1988), 362-371.

[3] Global perturbation of the Riemann problem for the system of one-dimensional issentropic flow, Partial Differential Equations (Proceedings, Tianjin 1986, edited by S.S.Chern), Lecture Notes in Mathematics 1306, Springer-Verlag, 1988, 131-140.

[4] Global discontinuous solutions to a class of discontinuous initial value problems for the system of isentropic flow and applications, Chin. Ann. of Math., 10B(1989), 1-18.

[5] Global classical solutions to typical free boundary problems for quasilinear hyperbolic systems, Science in China, Series A, 33(1990), 769-783.

[6] Solutions discontinues globales du problème de Riemann généralisé pour les systèmes hyperboliques quasi linéaires de lois de conservation, C. R. Acad. Sci. Paris, 310, Série 1(1990), 111-114.

[7] Global existence of classical solutions to the typical free boundary problem for general quasilinear hyperbolic systems and its applications, Chin. Ann. of Math., 11B(1990), 15-32.

[8] Global shock solutions to a class of piston problems for the system of one dimensional isentropic flow, Chin. Ann. of Math., 12B(1991), 495-499.

A.Majda

[1] The existence of multi-dimensional shock fronts, Memoirs of the American Mathematical Society, Number 281, 1983.

B.Riemann

[1] Über die Fortpflanzung ebener Luftwellen von endlicher Schwingungsweite, Abhandlungen der Gesellschaft der Wissenschaften zu Göttingen, Mathematisch-physikalische Klasse 8, 43(1860).

A.N.Tikhonov and A.A.Samarsky

[1] Discontinuous solutions of first order quasilinear equation (in Russian), ДОКЛАДЫ АКАДЕМИИ НАУК СССР , 99(1954), 27-30.

Wang Rou-huai and Wu Zhuo-qun

[1] Existence and uniqueness of solutions for some mixed initial boundary value problems of quasilinear hyperbolic systems in two independent variables (in Chinese), Acta Scientiarum Naturalium of Jilin University, 2(1963), 459-502.

Ying Long-an and Wang Ching-hua
[1] The discontinuous initial value problem of a reacting gas flow system, Trans. Amer. Math. Soc., 266(1981), 361-387.

Zhou Xiao-lin
[1] The centered wave solution with a large amplitude for quasilinear hyperbolic systems, Northeastern Mathematical Journal, 3(1987), 439-451.
[2] Local solutions of the large amplitude centered wave problem for quasilinear hyperbolic systems, Chin. Ann. of Math., 10A(1989), 119-138.

Minimal Surfaces in Riemannian Manifolds[†]

Guangyin Wang[††]

Institute of Mathematics, Academia Sinica, Beijing

In [JW] the disk-type minimal surfaces spanning a given Jordan curve Γ in a Riemannian manifold N is studied. In 1948, C.B. Morrey [M] proved that there exists at least one solution to this problem. In fact he proved the existence of the surface of least area among the surfaces with boundary Γ. An interesting question is whether there are other minimal surfaces with the same boundary besides. When the ambient space N is the standard n-sphere S^n, M.Ji and G.Y. Wang contributes a neat result:

Conclusion A. Each smooth Jordan curve Γ in S^n bounds at least two minimal surfaces, sometimes infinitely many ones.

It is well-known that, in Euclidean space there exists only one geodesic joining each pair of points, but in a curved manifold, in contrast, there may exist many geodesics when the manifold has a rich topology. This, in fact, is the content of Ljusternik-Schnirelmann theory or Morse theory (cf. [B]). Then it is natural to ask if similar results hold for minimal surfaces. There have been indeed a few works([CE], [St], [Tr] etc.) attempted this problem. By overcoming the well-known difficulties, M.Ji and G.Y. Wang succeeded in establishing a sound topological index theory to characterize the structure of the set of all co-boundary minimal disks. Conclusion A is simply a test of their theory. An outline of their work is the following:

The target manifold N is assumed to be embedded isometrically in $\mathbb{R}^k$ for some k without loss of generality. Let

$$W_\Gamma^{1,p} = \{u \in W^{1,p}(D, \mathbb{R}^k); u(D) \subset N, u(\partial D) \subset \Gamma, \deg(u|_{\partial D}) = 1\},$$

$$X = \{u \in W_\Gamma^{1,p}, u|_{\partial D} \text{ is weakly monotone}\},$$

where D is the unit disk: $\{(x, y) \in \mathbb{R}^2; x^2 + y^2 < 1\}$, and $p > 2$ to be fixed later.

It can be proved that $W_\Gamma^{1,p}$ is a Finsler manifold with the Finsler structure induced by $W^{1,p}(D, \mathbb{R}^k)$, and the closed subset X of it, though without a manifold structure, is locally convex. Thus with respect to X a critical point theory is feasible for compact functionals. They proved

---

† Supported by Foundation of Academia Sinica and NNSF of China

†† 1991 Mathematics Subject Classification: 58E12, 49Q05, 53A10, 53C42

C. Gu et al. (eds.), Partial Differential Equations in China, 104–110.

© 1994 Kluwer Academic Publishers.

Theorem 1. If u is critical for the energy functional

$$E(u) = \frac{1}{2} \int_D |\nabla u|^2 dxdy, \ \forall u \in X$$

with respect to X, then u is a conformal harmonic parametrization of a minimal surface.

Since the energy functional is not compact, the critical point theory in being can not be applied directly. To overcome the lack of compactness, they introduce the perturbed functionals according to [Uh]: for $\varepsilon > 0$,

$$F^\varepsilon(u) = \int_D (\frac{1}{2}|\nabla u|^2 + \frac{\varepsilon}{p}|\nabla u|^p) dxdy, \ u \in X,$$

which is proved to satisfy Palais-Smale condition with respect to X. Then the difficulty of lack of compactness turns to an a priori estimate, independent of $\varepsilon(\varepsilon > 0)$, for the set of critical points u_ε with respect to X. Denote

$$m^* = \inf\{E(u); u \in X\},$$

$$s_0 = s_0(N) = \inf\{E(u); u : S^2 \to N \text{ harmonic nonconstant}\}.$$

They proved

Theorem 2. There exists $p_0 > 2$ with the property that for $b < m^* + s_0$ and $B > 0$ there exists $\alpha = \alpha(b, B) > 0$ such that if u is critical for $F^\varepsilon(\varepsilon > 0, 2 < p \le p_0)$ with $F^\varepsilon(u) \le B$ and $E(u) < b$, then $\|u\|_{2,2} \le \alpha$.

Remark. Theorem 2 offers an analysis foundation of the study of minimal surfaces. The estimate there cannot hold when $\varepsilon = 0$ because the energy functional admits Möbius group which is noncompact. Thus this estimate which is uniform in ε is by no means evident. For the proof, the troubles arise mainly from that the boundary value $u_\varepsilon|_{\partial D}$ is possibly divergent as $\varepsilon \to 0$. For this reason, they choose a sequence of conformal transforms φ_ε of the unit disk such that $v_\varepsilon = u_\varepsilon \circ \varphi_\varepsilon$ satisfy the "three-point-condition", and then study $\{v_\varepsilon\}$ instead of $\{u_\varepsilon\}$. By doing so, the sequence of boundary values $\{v_\varepsilon|_{\partial D}\}$ becomes equicontinuous according to Courant-Lebesque-Lemma, but v_ε is no longer critical for the former functional $F^{\prime\varepsilon}$, in fact is critical for the new functional

$$H_\varepsilon(u) = \int_D (\frac{1}{2}|\nabla u|^2 + \frac{\varepsilon}{p}g_\varepsilon^{p-2}|\nabla u|^p) dxdy, \ \forall u \in X,$$

where the coefficient g_ε, depending on ε, diverges possibly as ε goes to zero because of the non-compactness of the Möbius group. At last the possibility of the divergence is successfully excluded by the following identity.

Theorem 3. For $\varepsilon > 0$, the critical point u of F^ε satisfies

$$\int_D x \cdot |\nabla u|^p dxdy = 0.$$

Notice that this identity may not hold when $\varepsilon = 0$. In addition, the "blow up" analysis , a common approach to get an a priori estimate, usually relies on some regularity theorem. They eventually got an available regularity theorem.

Theorem 4. If u is critical for functional

$$H(u) = \int_D (\frac{1}{2}|\nabla u|^2 + \frac{1}{p}g^{p-2}|\nabla u|^p)\,dx\,dy, \forall u \in X,$$

with respect to X, where $g = 0$ or $g(x, y) \in C^1$ and $g > 0$, then $u \in W^{2,2}$. Moreover, $\|u\|_{2,2} \le c$ depends only on $H(u)$, $\|u\|_{1,p}$, $\|\nabla(\log g)\|_{L^\infty}$ and p.

Remark. While interior regularity is by now a trivial matter, the regularity up to the boundary is not easy to build for the problem considered here. Since X is not a Banach manifold, the corresponding boundary conditions satisfied by the critical points of H with respect to X, is an alternative:

$$\frac{\partial u}{\partial r} \cdot \frac{\partial u}{\partial \theta} = 0 \quad on \quad \partial D,$$

in smooth setting, where (r, θ) being polar coordinates. On such conditions even some routine methods, such as the standard localization, do not work. Thus some new techniques are needed for the proof. Theorem 4, in fact, has its independent interest in theory of partial differential equations.

Based on the above theorems, Morse-Tompkins-Shiffman's theorem can be extended conveniently to manifold. They proved the following theorem stated by S.T. Yau in [Y].

Conclusion B.   Suppose that the compact manifold N admits no minimal sphere. If there are two strictly stable minimal disks bounded by the Jordan curve Γ, then there exists another minimal surface bounded by Γ.

In fact what they got was a stronger result.

Conclusion C. If there are two strictly minimal disks u_1 and u_2 bounded by the Jordan curve Γ where u_1 and u_2 are their conformal parametrizations respectively, then there exists another minimal surface bounded by Γ provided that u_1 and u_2 can be joined by a path $\gamma \in C^0([0, 1], X)$ such that

$$\max_{t \in [0,1]} E(\gamma(t)) < m^* + s_0.$$

Conclusion B is an evident consequence of Conclusion C, because the assumption that N admits no minimal sphere means $s_0 = \infty$.

Based on the analysis foundation,   both Morse theory and Ljusternik–Schnirelmann theory can be established for energy functional E on the space X by perturbed method (cf [Uh]). However Morse theory just gives information only when Γ bounds non-degenerate minimal surface. And Ljusternik–Schnirelmann theory also is unsatisfactory, in some sense, for the study of minimal surfaces. As we know that a single minimal disk is characterized by a whole critical orbit of E, yielded by the conformal group of D, whose category is but two in sense of Ljusternik-Schnirelmann, the notion of category gives hardly precise information about the set of minimal surfaces. M.Ji. and G.Y. Wang found a new topological index such that the index of an orbit is reduced to one and the index of X is kept nontrivial, and then established an appropriate index theory. Here is the definitionn of it. Denote

$$[u] = \{u \circ e^{i\tau}; \tau \in [0, 2\pi)\} \ \forall u \in W_\Gamma^{1,p}.$$

Definition. For $A \subset W_\Gamma^{1,p}$, define the index of A as $i(A) = \inf\{k > 0;$ there exists k closed set $B_1, B_2, \cdots, B_k$ such that $A \subset \bigcup_{j=1}^{k} B_j$ and B_j is deformable into $[u_j]$ for some $u_j \in W_\Gamma^{1,p}$ where $j = 1, 2, \cdots, k\}$.

Here a set B is called to be deformable into some $[u]$ if there exists a continuous map $h : [0, 1] \times B \to W_\Gamma^{1,p}$ with $h(0, \cdot) = id.; h(1, B) \subset [u]$.

With such an index, they attained the following minimax principle (a multiple solution theorem) for minimal surfaces. Let Z^0 be a connected component of $W_\Gamma^{1,p}$ and $X^0 = X \cap Z^0$. Let

$$c_j = c_j(E) = \inf\{\sup_A E; A \subset X, i(A) \geq j\} \text{ for } 1 \leq j \leq i(X^0).$$

Theorem 5. With the above notation, the following statements hold for $j = 1, 2, \cdots, i(X^0)$: (1) If $c_j < m^* + s_0$, then c_j is a critical value of E with respect to X; (2) if $c_j = c_{j+1} = \cdots = c_{j+m} = c < m^* + s_0$ and c is not a limit of critical values of E, then $i(K_c) \geq m + 1$, where K_c is the set of critical points with energy c.

Corollary: If there is an integer $j \in [1, i(X^0)]$ such that $c_j < m^* + s_0$, then there exist at least j minimal surfaces in N with boundary Γ.

As a test of the topological index theory, Conclusion A is obtained, which is about disk type(simply connected) minimal surfaces–the Plateau problem. Notice that in Conclusion A no restriction (of course except for smoothness) is imposed on boundary Γ. However, for the annulus type minimal surface–the Douglas problem, situation is quite different. Even in the Euclidean space $\mathbb{R}^3$, for a pair of contours Γ_1 and Γ_2, even possibly there does not exist one minimal surface bounded by $\Gamma_1 \bigcup \Gamma_2$. A condition–the socalled Douglas condition–was introduced for the existence of a solution. Extended the Donglas condition, M.Ji established a multiple solution theory for minimal annuli co-boundaries in Riemannian manifolds. We give a brief review as follows.

A minimal annulus spanning Γ_1 and Γ_2 is a conformal harmonic map from an annulus $A_\rho = \{(x, y) \in \mathbb{R}^2, \rho < \sqrt{x^2 + y^2} < 1\}$ to the ambient manifold N with $u(c_\rho) = \Gamma_1$ and $u(\partial D) = \Gamma_2$, where $c_\rho = \{(x, y) \in \mathbb{R}^2; x^2 + y^2 = \rho^2\}$. A map

$u : A_\rho \to N$ is harmonic if it is critical for the energy functional

$$E(u; A_\rho) = \frac{1}{2} \int_{A_\rho} |\nabla u|^2 dx dy$$

with respect to interior variation of u, and in order to make it be simultaneously conformal, it should also be critical with respect to both boundary variation of u and variation of inner redius ρ of domain A_ρ. M.Ji performs this complicated variations in a varied way. Now we introduce some spaces of maps. Let $p > 2, 0 < \rho_0 < 1$, define

$$W^{1,p}_{\Gamma_1,\Gamma_2} = \{u \in W^{1,p}(A_\rho, N); u(c_\rho) = \Gamma_1, u(\partial D) = \Gamma_2,$$
$$\deg(u|_{c_\rho}) = \deg(u|_{\partial D}) = 1\}$$

and

$$X = \{u \in W^{1,2}_{\Gamma_1,\Gamma_2}(A_{\rho_0}, N); u|_{c_{\rho_0}} \text{ and } u|_{\partial D} \text{ weakly monotone }\}.$$

The working space will be $X \times (0,1)$ which not only has no manifold structure but also is non-complete. By building some a priori estimates, she was able to establish an available critical point theory for the functional $F(u, \rho) = E(u^\rho; A_\rho)$ on the "bad" space $X \times (0,1)$, where $u^\rho(r, \theta) = u\left(\frac{\rho_0 - \rho + (1-\rho_0)r}{1-\rho}, \theta\right) \in W^{1,p}_{\Gamma_1,\Gamma_2}(A_\rho, N)$. She proved [J1] that

Theorem 6. If (u, ρ) is a critical point of F with respect to $X \times (0,1)$, then u^ρ, defined above, will be harmonic and conformal in A_ρ, therefore it characterizes a minimal surface of annulus type spanning Γ_1 and Γ_2.

Now we introduce some notations. Denote

$$Y_i = \{\varphi \in C^0(S_1, \Gamma_i); \ \varphi \text{ is weakly monotone}\}, \quad i = 1, 2,$$

$$W^{1,2}_\varphi = \{u \in W^{1,2}(D, \mathbb{R}^k); u(D) \subset N, a.e., u|_{\partial D} = \varphi, a.e.\},$$

$$W^{1,2}_{\varphi,\psi} = \{u \in W^{1,2}(A_{\rho_0}, \mathbb{R}^k); u(A_{\rho_0}) \subset N, a.e., u(\rho_0, \theta) = \varphi(\theta),$$
$$u(1, \theta) = \psi(\theta), a.e.\}.$$

Define

$$X_2(\Gamma_1, \Gamma_2) = \bigcup_{\varphi \in Y_1} \bigcup_{\psi \in Y_2} W^{1,2}_{\varphi,\psi};$$

$$X_2(\Gamma_i) = \bigcup_{\varphi \in Y_i} W^{1,2}_\varphi, \quad i = 1, 2.$$

Set

$$m^* = \inf\{F(u, \rho); (u, \rho) \in X_2(\Gamma_1, \Gamma_2) \times (0,1)\};$$
$$m_i = \inf\{E(u); u \in X_2(\Gamma_i)\}, \quad i = 1, 2,$$
$$d^* = m_1 + m_2.$$

For $A \subset X \times (0,1)$, set

$$i(A) = m$$

if A can be covered by m closed subsets $B_1, B_2, \cdots, B_m$ in $X \times (0, 1)$, with B_j deformable to $[u_j] \times \{\rho_j\}$ in $X \times (0, 1)$ for some $u_j \in X, \rho_j \in (0, 1)$ where $[u_j] = \{u_j(ze^{i\theta}); 0 < \theta \leq 2\pi\}$, and m is minimal with this property. And set

$$c_l = \inf\{\sup_A F; A \subset X \times (0, 1), i(A) \geq l\}, 0 \leq l \leq i(X \times (0, 1)),$$

where l are integers.

With these notations, she get a multiple solution theorem.

Theorem 7. If

$$c_l = c_{l+1} = \cdots = c_{l+m} = c < \min\{m^* + s_0, d^*\}$$

where $m \geq 0$, $1 \leq l \leq i(X \times (0, 1))$, and c is not a limit of the critical values of F, then $i(K_c) \geq m + 1$ where K_c is the set of critical points of F with value c.

Because a critical point of F w.r.t $X \times (0, 1)$ characterizes a minimal annulus bounded by Γ_1 and Γ_2, and each critical orbit is of index one, then Theorem 7 has the following

Conclusion D. If for some integer $l \in [1, i(X \times (0, 1))]$,

$$c_l < \min\{m^* + s_0, d^*\} \tag{*}$$

then there exist at least l minimal annuli spanning Γ_1 and Γ_2 in N.

Remark. The inequality (*) actually means two conditions: $c_l < m^* + s_0$ and $c_l < d^*$. When N is nonpositively curved, the former always holds (since $s_0 = +\infty$), while the latter may not hold if only Γ_1 and Γ_2 are far sufficiently apart. Thus the latter may be considered as a restriction on the "relative distance" between Γ_1 and Γ_2. Noting that $m^* = c_1 \leq c_l$, it implies the Douglas condition $m^* < d^*$, thus Conclusion D contains Morrey's result (cf.[M]).

Applying the general theory to the case $N = S^n$, M.Ji got a very simple criterion of existing of two minimal annuli co-boundaries.

Conclusion E. In S^n, for each pair of contours Γ_1 and Γ_2 satisfying the condition

$$c_2 < d^*, \tag{†}$$

there exist at least two minimal annuli bounded by Γ_1 and Γ_2.

It should be illustrated that what pairs of contours Γ_1 and Γ_2 will satisfy the criterion (†) and then they bound at least two minimal annuli in S^n.

A famous example, due to Lawson (cf. [L]), is as follows. In the 3-sphere

$$S^3 = \{(v, w) \in \mathbb{C}^2; |v|^2 + |w|^2 = 1\},$$

$$\sigma_1 := S^3 \cap \{(0, w) \in \mathbb{C}^2\} \text{ and } \sigma_2 := S^3 \cap \{(v, 0) \in \mathbb{C}^2\}$$

are great circles that are linked and of a constant spherical distance $\pi/2$ apart. For each $\theta \in [0, 2\pi)$,

$$\sigma_\theta := S^3 \cap \{(v, w); \arg(v\bar{w}) = \theta\}$$

is an embedded minimal annulus in S^3 with boundaries σ_1 and σ_2.

By a continuation argument, M.Ji got the following

Conclusion F. In the sphere S^3, let Γ_1 and Γ_2 be a pair of contours near σ_1 and σ_2 respectively, then there exist at least two minimal annuli in S^3 bounded by Γ_1 and Γ_2.

Recently, M. Ji proved more concise results.

Conclusion G. Suppose that Γ is a smooth Jordan curve which is linked with a great circle σ in S^3. Then Γ, σ bound at least two minimal annuli in S^3.

Conclusion H. Let Γ, σ be two smooth disjoint Jordan curves in S^n. Suppose that σ is a great circle which intersects a least area surface spanning Γ. Then Γ, σ bound at least two minimal annuli in S^n.

References

[B]   R.Bott, Lectures on Morse theory, old and new, Bull AMS, 7(1982), 331-358.

[CE] K.Chang and J.Eells, Unstable Minimal Surfaces Co-boundaries, Acta Mathematica Sinica, New Series (1986) Vol.2, No.3, 233-247.

[J1] M.Ji, An apriori estimate for Douglas problem in Riemannian manifolds, Acta Math. Sinica, New Series (1989) Vol.5, No.3, 235-249.

[J2] M.Ji, A remark on perturbation methods, Acta Math. Sinica,(1989) Vol.32, No.5, 628-631.

[J3] M.Ji, Minimal annuli in Riemannian manifolds, Acta Math. Sinica, New Series (1993) Vol. 9, No.1.

[JW] M.Ji and G.Y.Wang, Minimal surfaces in Riemannian manifolds, Memoirs AMS, 495 (1993).

[L]   H.B. Lawson, Complete minimal surfaces in S^3, Ann. Math., 92(1970), 335-374.

[M]   C.B.Morrey, Multiple integrals in the calculus of variations, Groundlehren Band, Springer, 130(1960).

[St] M.Struwe, On a critical point theory for minimal surfaces spanning a wire in $I\!R^n$, J.Reine Angew, Math. 349(1984), 1-23.

[Tr] A.J.Tromba, On the number of simply conncted minimal surfaces spanning a curve, Memoirs AMS 194(1977).

[Uh] K.Uhlembeck, Morse theory by perturbation methods with applications to harmonic maps, Trans. AMS 267(1981).

[Y]   S.T.Yau, Survey on partial differential equations in differential geometry, Seminar on Differential Geometry, Princeton University Press, 1982.

Microlocal Analysis[†]

Rouhuai Wang

Department of Mathematics, Jilin University

Shuxing Chen

Institute of Mathematics, Fudan University

Microlocal analysis is a branch formed in the middle sixties of the century. It has a history lasting only over twenty years. In the sixties and the beginning of seventies there were some Chinese mathematicians who noticed the formation and development of this branch. For instance, in the eve of the birth of pseudodifferential operators, the Fouier multiplier had been applied to establish the L^p theory of elliptic and parabolic boundary value problems in China. However, the systematic study began only after the late seventies. In this article we will introduce the results on the theory of microlocal analysis and its applications, which were obtained by Chinese mathematicians in the recent ten years. The introduction consists of three aspects. In the first section we introduce the results on the boundedness of pseudodifferential operators and Fourier integral operators. In the second section we discuss the theory of linear microlocal analysis and related problems. Then the last section is devoted to the nonlinear microlocal analysis and singularity analysis for nonlinear equations. As for the study of equations of mixed type or Fuchian type, since they have been discussed in the other articles of this book, we will not repeat them here.

1 Pseudodifferential operators and Fourier integral operators

L. *Hömander* gave a class of symbols S^m, if function $p(x, \xi) \in S^m$, then a pseudodifferential operator

$$p(x, D)u(x) = \frac{1}{(2\pi)^n} \int e^{ix \cdot \xi} p(x, \xi) \hat{u}(\xi) d\xi \tag{1.1}$$

where $\hat{u}(\xi)$ represents the Fourier transformation of u(x), the operator p(x, D) is a linear bounded operator from L^2 to L^2. In [A7] L. Homander established a complete L^2 theory. In this paper he also proved that for $q \le 2 \le r$ and $m \le -n(\frac{1}{q} - \frac{1}{r})$

$$\|p(x, D)u\|_{L^r} \le C\|u\|_{L^q}$$

† 1991 Mathematics Subject Classification: 35A27, 35S05, 35S30, 35S50, 35H05, 58G15, 58F17

† Supported by NNSF of China

C. Gu et al. (eds.), Partial Differential Equations in China, 111–126.

© 1994 Kluwer Academic Publishers.

However, the corresponding inequality for $q \leq r \leq 2$ or $2 \leq q \leq r$ was unknown at that time. Chang Kunching solved the open peoblem, he proved in [CK1].

Theorem 1.1 If $p(x, \xi) \in S^m(\mathbf{R}_x^n \times \mathbf{R}_\xi^n)$, $1 < q \leq r < \infty$, $m \leq -n(\frac{1}{q} - \frac{1}{r})$. then

$$\|p(x, D)u\|_{L^r} \leq C\|u\|_{L^q} \qquad (1.2)$$

In this theorem the restrictions $q \leq 2$ and $r \geq 2$ are relieved. This result was obtained in 1974, it was probably the first result in the field of microlocal analysis in China.

Up to the end of seventies there were many works on boundedness of pseudodifferential operators. It was noticed that in order to obtain the boundedness of these operators in L^2 or L^p, the requirement for the smoothness of symbols can be reduced. Wang Rouhai and Li Chenzhang in their paper [WL1] developed the method of harmonic analysis established by Coifman and Meyer in [A4], and proved the sharp L^2 boundedness of pseudodifferential operator defined by using amplitude. On the other hand, they also established the sharp estimates for the differentiability of the symbols is one order lower. And these conditions could not be decreased any more as showed in their examples.

In [WL1] the following symbol class is introduced: let λ and μ be non-negative integers, m be real, $0 \leq \delta \leq \rho \leq 1$, $0 \leq \epsilon \leq \rho \leq 1$, $\delta < 1$, $\epsilon \leq 1$, if

$$|D_x^\beta D_y^\gamma D_\xi^\alpha a(x, \xi, y)| \leq C < \xi >^{m-\rho|\alpha|+\delta|\beta|+\epsilon|\gamma|} \qquad (1.3)$$

is satisfied for all $|\alpha| \leq \lambda$ and $|\beta| + |\gamma| \leq \mu$, then $a(x, y, \xi)$ is said to belong to the class $S_{\rho,\lambda;\delta,\mu;\epsilon,\nu}^m$. Similarly, the symbol class $S_{\rho,\lambda;\delta,\mu}^m$ can be defined. The main result in [WL1] is (in the sequel we denote $k = [\frac{n}{2}] + 1$).

Theorem 1.2 If $a(x, \xi, y) \in S_{\rho,2k;\delta,k;\epsilon,k}^0$, then the pseudodifferential operator with the amplitude

$$a(x, D, y)u = \frac{1}{(2\pi)^n} \int \int e^{i(x-y)\xi} a(x, \xi, y)u(y)dyd\xi \qquad (1.4)$$

is L^2 bounded.

Theorem 1.3 If $a(x, \xi) \in S_{\rho,k;\delta,k}^{-m}$, $\rho > 0$, $0 \leq m \leq \frac{n}{2}(1 - \rho)$ and $0 \leq (\frac{1}{2} - \frac{1}{p})n(1 - \rho) \leq m$, then the pseudodifferential operator a(x, D) with the symbol $a(x, \xi)$ is L^p bounded.

Theorem 1.4 Assume $a(x, \xi, y) \in S_{\rho,2k;\delta,k;\epsilon,k}^{-m}$. The pseudodifferential operator with the amplitiude $a(x, \xi, y)$ is L^p bounded, if and only if $n(1 - \rho)|\frac{1}{2} - \frac{1}{p}| \leq m$.

In the papers [LC1],[LC2] some further results on the boundedness of pseudodifferential operators are obtained. Moreover, in [LC1] the boundedness in the space $H_p^{s/\alpha}(\mathbf{R}_+^n)$ is discussed, the result is

Theorem 1.5 Assume that $p(x, \xi)$ is a classical symbol of order m, satisfying the transmission condition (see [A8]), then for any $s > -a_n/p'$, the pseudodifferential operator p(x,D) is continuous from $H_p^{s/\alpha}(\mathbb{R}_+^n)$ to $H_p^{(s-m)/\alpha}(\mathbb{R}_+^n)$.

Qiu Qingjiu studied in [QQ1–QQ3] the boundedness of Fourier integral operators with the form

$$Fu(x) = \int e^{i\phi(x,\xi)} a(x, \xi) u(\xi) d\xi \qquad (1.5)$$

in the space L^p, H_p^s and $B_{p,q}^s$. The systematic study on the boundedness of Fourier integral operators is initiated by M. Beals (see [A1]). He presented some conditions on the phase function in (1.5) as
   $\phi(x, \xi)$ is analytic with respect to ξ in $\mathbb{R}^n \setminus \{0\}$,

   the set $S_x = \{\xi; \quad \xi \in \sup a(x, \xi), \quad \phi = 1\}$ is convex $\qquad (1.6)$
the map $S_x \ni \xi \to \nabla_\xi \phi(x, \xi)/|\nabla_\xi \phi(x, \xi)|$ is bijective.
   In [QQ2] Qiu held these conditions and proved

Theorem 1.6 Assume $a(x, \xi) \in S_{1,\delta}^{-m}$, $s \geq 0$, $\delta \leq 1$, then under the assumptions (1.6) the operator (1.5) is bounded from $(H_p^s)_{comp}$ to $(H_p^s)_{loc}$, and it is also bounded from $(B_{p,q}^s)_{comp}$ to $(B_{p,q}^s)_{loc}$ for $1 \leq q \leq \infty$.

In [QQ3] the condition (1.6) is alleviated to that the phase function $\phi(x, \xi)$ is nondegenerate and the rank of its Hessian is greater than r at the critical point. Then the operator (1.5) is bounded from $(L^p)_{comp}$ to $(L^p)_{loc}$, provided the amplitude belongs to $S_\rho^{-m}(\frac{1}{2} < \rho \leq 1)$ and $m \geq (n(3-p) - 2 - r)|\frac{1}{2} - \frac{1}{p}|$.

On the study of the theory of Fourier integral operators Wang Rouhwai paid more attention to its global theory. In [A8] Melin & *Sjöstrand* developed the theory of Fourier integral operators with complex phase. For the corresponding global theory the construction of Keller-Maslov line bundle is a key point. Wang Rouhwai & Cui Zhiyong in [W&C1] made the algebraic considerations necessary for the definition of the Keller-Maslov line bundle on a Lagrangian manifold in the complex case. They established an extension of Leray's formula in the case of real phase, and then gave a method of constructing Maslov line bundle. This work has brought to *Hörmander's* attention, and was referred in his famous book [A8].

2   The study on linear partial differential equations

The development of microlocal analysis greatly influenced the study of partial differential equations. Up to seventies the systematic theory on linear partial differential equations has been established. Different from the classical one this theory involves many new subjects including hypoellipticity, local solvability and singularity propagation etc. Since the study on the operators of principal type were almost complete in the early seventies, then people are mainly concerned with the operators with multiple characteristics. In this aspect Chinese mathematicians also made their contributions.

2.1 Hypoellipticity

Based on the works of *Höemander* [A7] and Rothschild & Stein [A12], Huang Yumin in [HY4] studied the hypoellipticity of operators with multiple characteristics. He extended the optimal estimates of internal regularity for the operators with the form "sum of squares" in [A7] to the case of pseudodifferential operators. Let p(x, D) is a classical pseudodifferential operator of order m in the domain $\Omega \subset \mathbb{R}^n$, $p_m(x, \xi) \geq 0$ and $Re p_{m-1}(x, \xi) = 0$. Let $Q_0(x, \xi)$ be the imagine part of the subprincipal symbol

$$Q_j(x, \xi) = \begin{cases} \dfrac{\partial p_m}{\partial \xi_j}(x, \xi), & j=1,2,..,n; \\[2mm] (1 + |\xi|^2)^{-1/2} \dfrac{\partial p_m}{\partial x_j}(x, \xi) & j=n+1,...,2n \end{cases}$$

Denote by Q_j the proper supported pseudodifferential operator with symbol $Q_j(x, \xi)$, $I = \alpha_1, ..., \alpha_k$, where $\alpha \in 0, 1, ..., 2n$, and denote

$$Q_I = [Q_{\alpha_1}, [Q_{\alpha_2}, ..., [Q_{\alpha_{k-1}}, Q_{\alpha_k}]...]$$

$$|I| = \sum_{j=1}^k \lambda_j, \quad \lambda_j = 1 \text{ if } \alpha_j \neq 0;\ 2 \text{ if } \alpha_j = 0$$

then the following theorem holds.

Theorem 2.1 If for any $(x, \xi) \in \Omega \times \mathbb{R}^n \setminus \{0\}$, there is $|I| \leq 3$, such that Q_I is elliptic at (x, ξ), then for any $s \in \mathbb{R}$

$$u \in (\Omega), \quad Pu \in H^s_{loc}(\Omega) \Rightarrow u \in H^{m+s-\frac{1}{3}}_{loc}(\Omega) \tag{2.1}$$

This theorem improved the corresponding result in [A8], the proof can be referred to [HY4].

Luo Xuebo and Fu Chuli also studied hypoelliptic operator in [L&F1], they introduced the concept of supplemental operator, and proved that every linear partial differential operator has supplemental operator of principal type. Then by using the result on the operators of proncipal type some operators of non-proncipal type are obtained. For instance, we have

Theorem 2.2 Let $p(x, D_x)$ be a linear partial differential operator in $\Omega_x \subset \mathbb{R}^n$ with order $m > 1$, $p(x, D_x)$ is hypoelliptic if the following conditions hold: i) Rang $p_m(x, \xi) \subset \Gamma \cup (-\Gamma)$, where $\Gamma \subset C$ is an angle region on the complex plain with its vertex at the origin; $\Gamma < \pi$. ii) There is $\alpha_0 \notin \Gamma \cup (-\Gamma)$, $\alpha_0 \neq 0$, such that for every $z \in C$ and $(x^0, \xi^0) \in T^*(\Omega_x) \backslash \{0\}$ satisfying $p_m(x^0, \xi^0) = 0$ and $|grad_\xi Re(zp_m)(x^0, \xi^0)| + |Re(z\alpha_0)| \neq 0$, the function $Im(zp_m(x, \xi))$, restricted to the bicharacteristic strip of $Re(zp_m(x, \xi))$ through (x^0, ξ^0), does not change sign at that point and does not vanish identically in any neighborhood of (x^0, ξ^0) on this bicharacteristic strip.

Hong Jiaxing investigated hypoellipticity of some differential operators of second order by using the microlocal equivalent transformation of operators which multiple characteristics (see [HJ1], [HJ2]). Let p be the symbol of a pseudodifferential operator of second order, $p \sim p_2 + p_1 + ...$, where $p_2 = t\tau^2 + e(x, t\xi, \tau)$ and e is a real homogeneous function of degree 2, satifying $e(x, t, 0, \tau) = 0$, $\frac{\partial e}{\partial \xi_j}(x, 0, 0, \tau) = 0$

(j=1,...,n-1). Obviously, in this case the Hamiltonian vector field H_{p_2} of p_2 on $\Sigma = \{t = 0, \xi_1 = \xi_2 = ... = \xi_{n-1} = 0\}$ is parallel to $\xi \frac{\partial}{\partial \xi} + \tau \frac{\partial}{\partial \tau}$.

Theorem 2.3 If p satisfies the above assumptions, then for any $(x_0, 0, 0, \tau^0) \in \Sigma$, there exists one-dimensional neighborhood $\Gamma_1(x_0, 0, 0, \tau^0) \subset T^*(\mathbb{R}^n_{x,t})$, and $\Gamma_2(0, 0, 0, \tau^0) \subset T^*(\mathbb{R}^n_y)$, a classical homogeneous transformation $\chi : \Gamma_1 \ni (x, t, \xi, \tau) \to (y_1, ..., y_n, \eta_1, ..., \eta_n) \in \Gamma_2$ and a corresponding Fourier integral operator F, such that

$$FPF^{-1} \sim A_1(y_n D_n + q(y'))A_0$$

where A_0, A_1 are microlocally elliptic operators at $(0, 0, 0, \tau^0)$ and $q(y') = p_1/\tau)|_{\kappa-1} + \sqrt{-1}$ mod (real part)

This theorem can be applied to discussing the hypoellipticity of the operator $Lu = \sum a^{ij} u_{x_i x_j} + \sum b^i u_{x_i} + cu$. In [15] the following result is obtained.

Theorem 2.4 If for the operator Lu, there is surface S: $\phi(x) = 0$, such that the following conditions hold on S:

i) $a^{ij}\phi_{x_j} = 0$, $d(a^{ij}, \phi_{x_i}, \phi_{x_j}) \neq 0$.

ii) (a^{ij}) has (n-1) positive characteristic roots.

iii) $Re(b^i - a^{ij}_{x_j})\phi_{x_i} + \frac{1}{2} a^{ij}_{x_s} \phi_{x_i} \phi_{x_j} \phi_{x_s} |\nabla \phi|^{-2} > 0$ on S.

then for any neighborhood $O(x_0)$ of x_0, there exists $O_1(x_0) \subset\subset O(x_0)$, such that $u \in L^2(O(x_0))$, $Lu \in C^\infty(O(x_0))$ implies $u \in C^\infty(O_1(x_0))$.

Xu Chaojiang studied subelliptic eatimates for partial differential operators of second order with non-smooth coefficients. His results can be applied to discuss the regularity of solutions of degenerate elliptic equations. By using the concept of subunit phere $B_L(x, \delta)$ corresponding to a given operator L, introduced by Fefferman and Phong (see [A5]), He proved in [XC3]

Theorem 2.5 Assume that for any $K \subset\subset \Omega$, there exist constants $C > 0$ and $\epsilon > 0$, such that

$$B_{-\Delta}(x, \rho) \subset B_L(x, C\rho^\epsilon), \quad \forall x \in K, \quad \forall \rho \in (0, 1)$$

then for any $K \subset\subset \Omega$, there exist constants $C' > 0$ and $\sigma > 0$, such that

$$\|\phi\|_\sigma^2 \le C'\{Re < L\phi, \phi > + \|\phi\|_0^2\}, \quad \forall \phi \in C_0^\infty(K) \tag{2.2}$$

In [X&Z1] the smoothness up to boundary for solutions of nonlinear and nonelliptic Dirichlet problem is also considered.

2.2 Local solvability

Huang Yinmin studied in [HY2] estimates of pseudodifferential operators of non-principal type and then by using his estimate to establish some corresponding result on local solvability. Let $p(x,D)$ be a classical pseudodifferential operator of order m defined on an open set in $\mathbb{R}^n$ with real symbol. Denote by $q(x, \xi)$ its sub-principal symbol $p_{m-1}(x, \xi) - \frac{1}{2i} \sum_{j=1}^n \frac{\partial^2 p_m}{\partial x_j \partial \xi_j}(x, \xi)$, and

$$\Theta = \{\xi \in \mathbb{R}^n \setminus \{0\}; p_m(x_0, \xi) = 0, \nabla_\xi p_m(x_0, \xi) = 0, \nabla_x p_m(x_0, \xi) \| \xi\}$$

$$\alpha(\xi) = |\xi|^2 \sum_{j=1}^n \xi_j \frac{\partial p_m}{\partial x_j}(x_0, \xi), \quad \forall \xi \in \Theta$$

Then the following theorem holds:

Theorem 2.6 Assume that

i) If $\xi \in (\mathbb{R}^n \setminus \{0\}) \setminus \Theta$ and $p_m(x_0, \xi) = 0$, then the projection of bicharacteristic strip through (x_0, ξ) on the base space does not stay at x.

ii) There is a real number l, such that $Im q(x_0, \xi) - l\alpha(\xi) \ne 0, \quad \forall \xi \in \Theta$

Then there exists a real number s and a neighborhood U of x, such that for any real t, the estimate

$$\|u\|_{s+m-1} \le C(\|Pu\|_s + \|u\|_t), \quad u \in C_0^\infty(U) \tag{2.3}$$

holds.

This estimate implies some results on local solvability of non-principal type, see [HY2]. The study on local solvability in China can also refer to [DX1], [QQ2], [WL1] etc. Besides, Lu Lijiang studied solvability of Cauchy problem of a class of pseudodifferential operator in [LLj1], Luo Xuebo studied solvability in the space.

Chen Shuxing discussed necessary conditions of local solvability of boundary value problems for linear partial differential equations in [CS9], where the concept of overcovering of polynomials is introduced: Let $r(t)$ be a polynomial of degree m, and $p_j(t)_{1 \le j \le k}$ a system of polynomials If $k < m$, and there are polynomials $p_{j1}(t), ..., p_{jm}(t)$, which are linearly independent modulo $r(t)$, then $p_j(t)$ is called overcovering to $r(t)$.

Theorem 2.7 Let $\Omega \subset \mathbb{R}^n$ be a neighborhood of x_0, P be a linear partial differential operator of order m with C^∞ coefficients, $\Gamma = \Omega \cap \{x_n = x_{0n}\}$ be non-characteristic with respect to P, B_j (j=1,...,) be operators of order m given on Γ. Then the boundary value problem

$$Pu = f, \quad in \quad \Omega \cap \{x_n > x_{0n}\}$$

$$B_1 u = g_1, ..., B_l u = g_l, \quad on \quad \Omega_0 \tag{2.4}$$

is local solvable at x_0, if for any $\xi' \in T^*_{x_0}(\Omega)$, the system b_j of polynomials is not an overcovering to $p_+ p_h$, where b_j is the symbol of B_j, $p_+ p_h$ are the product of all factors with real roots or complex roots with negative imaginary part in the decomposition of symbol $p(x, \xi)$.

For a given partial differential operators with coefficients involving a parameter, the continuous change of the parameter may cause discontinuous change of solvablility, uniqueness or other properties of this operator, this is called discrete phenomena. The appearance of eiginvalues for elliptic operators is one of such phenomena. F. Treves [A13] investigated such phenomena for an operator with multiple charateristics. Chinese mathematicians Wang Guangyin, Mai Mingcheng, Lu Zhujia, Wang Chungfang and Zheng Xilin also touched this topic. They mainly studied the discrete phenomena of various problem for operator

$$u_{xx} - x^2 u_{tt} - p u_t = 0 \tag{2.5}$$

For instance, [WML1] indicated that for the Cauchy problem and Goursat problem of (2.5), the solution is unique if $p \neq 1, 3, 5, ..$, but in the case p=1,3,5,... the solution will be unique only after adding some consistency conditions at x=0.

2.3 Propagation of singularities

The problem on propagation of singularities for operators with multiple characteristics is more complicated than the problem for principal operators. Here we would like to mention a few results of this topics. Wu Fangton in [WF2] studied the problem on reflection of singularities at diffractive point on boundary for operators with constant multiplicity of characteristics. He indicated that near boundary the singularity of u reflects and propagates along bicharacteristic strip at diffractive point for such operator P, provided Pu is C^∞. Qiu Qingjiu and Hong Jiaxing considered reflecting and unreflecting phenomena of singularities near boundary for a class of degenerate hyperbolic operators. In [Q&Q1] the authors proceeded the singularity analysis for a non−effective hyperbolic operator. They found that for such operators singularity may stop propagating at some multiple characteristic point, moreover, as an example they constructed a solution of the operator $\partial_t^2 - t^{2l} \sum_{j=1}^n D_{x_j}^2 U - i a t^{l-1} \Lambda$ by Fourier integral operators. This example shew the unreflecting phenomena do exist.

In [HJ5] the operator

$$Pu = D_t^2 u - t^{2N+1} \sum_{j=1}^n D_{x_j}^2 u + \sum_{j=1}^n p_j D_j u + qu \tag{2.6}$$

is discussed. Let $p_{\bar{j}} = t^{s_j} \tilde{p}_j(x,t)$ satisfying $\tilde{p}_j(x_0,0) \neq 0$, and s(p)=min s. Then the unreflecting phenomena may appear if $s(p) \leq N - 1$, it means that the terms of lower order degenerate slowly near t=0.

2.4 Application of microlocal analysis to *Schrödinger* operators

Wang Xueping successively applied microlocal analysis to some problems related to *Schrödinger* operator, mainly the problems on semi-classical approximation of evolution equations and time-delay of scattering theory. He obtained a series of good results.

In [WX4] and [WX6] Wang considered the equation

$$ih\frac{\partial}{\partial t} = A(t)u(t) \tag{2.7}$$

where A(t)=a(x, hD; t) is a family of pseudodifferential operators depending on t, and h is a small parameter. The author constructed the h-parametrix of (2.7) and established a semi-classical version of Egorov's theorem. By using these results, he proved that the limit of the solution of (2.7) converges to the solution of the classical Hamiltonian system as $h \to 0$ and then gave a regorous proof of corresponding principle in quantum mechanics and classical mechanics.

In [WX8] and [WX10] Wang took the operator A(t) in (2.7) as $-h^2\Delta + V$ and constructed a global approximation of the solution of *Schrödinger* equation. He obtained a uniform decay rate of the Unitary group $e^{ih^{-1}\xi A h}$ with respect to t and h in the weighted L^2 space, and then proved the uniform decay rate of energy in quantum mechanics is equivalent to the non-trapping condition in classical mechanics.

In [WX5], [WX7] and [WX9] the author proceeded detailed study for the time-delay problem from many aspects (existence, continuity, expression, the relation among scattering data and semi-classical approximation). He proved the existence of time-delay operator in the scattering theory with short range potential. Before his work there is only corresponding result on existence of time-delay operator when V(x) is spherical symmetric and $V(x) = O(|x|^{-2-\epsilon})$ as $|x| \to \infty$. Since in the case when V(x) delays slower than $O(|x|^{-1})$, the divergence of some integrals would cause trouble in proving the existence of the time-delay operator, then in [WX9] and [WX10] the author established careful estimates for a class of pseudodifferential operators with parameter by microlocalization along the trajectory on which the particle is moving. The details can be found in [WX10] etc.

3 Nonlinear microlocal analysis

Since a series of successful results in the study on linear microlocal analysis has been obtained, some mathematicians started to turn their attention to the study on nonlinear problems. A remarkable example is the establishment of the theory of paradifferential operators by French mathematician J.M. Bony and the successful application to the propagation of singularities of solutions to general nonlinear equations (see [A3]). In China the situation is similar. The basic theory of paradifferntial operators and its first applications have been introduced in detail in the book "Introduction to paradifferntial operators" written by Chen Shuxing, Qiu Qingjiu and Li Chengzhang [CQL1].

3.1 Nonlinear subelliptic operators

Xu Chaojiang studied the nonlinear equations of second order with the form

$$F(x, u, \nabla u, \nabla^2 u) = 0 \tag{3.1}$$

where is C^∞ with respect to its arguments. Assume that $u \in C_{loc}^\rho(\Omega)$ with $\rho \geq 4$ is a solution of (3.1) in the domain $\Omega \subset \mathbb{R}^n$. The corresponding linearized operator is

$$L = \sum_{i,j=1}^n a_{ij}(x)\partial_i\partial_j + \sum_{j=1}^n b_j(x)\partial_j + c(x) \tag{3.2}$$

with $a_{ij}, b_j, c \in C_{loc}^{\rho-2}(\Omega)$. The operator L is called degenerate elliptic if $(a_{ij}(x)) \geq 0$ for any $x \in \Omega$, and L is called subelliptic if for any $K \subset \Omega$, there are constants $\epsilon > 0$ and $C > 0$, such that

$$\|\phi\|_s^2 \leq C\{|<L\phi, \phi>| + \|\phi\|_0^2\}, \quad \forall \phi \in C_0^\infty(K) \tag{3.3}$$

In [XC5] Xu proved.

Theorem 3.1 If $u \in C_{loc}^\rho(\Omega)$, $\rho \geq 4$ is a solution of (3.1), and the linearized operator is subelliptic, then $u \in C^\infty(\Omega)$.

The proof of this theorem relies on the theory of paradifferential operators. The author also gave some improvement of Bony's theorem in his paper in order to set up a priori estimate for the solution with regularity as lower as possible.

Furthermore, in [X&Z1] the authors discussed regularity of the solutions to Dirichlet problem for (3.1). Denote by $L_0(x, \xi)$ the principal symbol $L_0^{(j)}(x, \xi) = (\frac{\partial L_0}{\partial \xi_j})(x, \xi)$, then we have

Theorem 3.2 Let r be an integer and u belong to $C^\rho(\Omega)$ with $\rho > max(5, r+3)$. Suppose that

i) $L_0(x, \xi) \geq 0$ for all (x, ξ).

ii) it denotes the set of brackets of vector fields $L_0^{(j)}$ of order less than or equal to r, then at each point of Ω one can find n elements in which are linearly independent;

iii) $\partial\Omega$ is noncharacteristic for $L_0(x, D)$; then u belong to $C^\infty(\Omega)$. Hong Jiaxing and Zuily discussed the regularity of solution of Monge–Ampere equation. Because the result has been briefly introduced in another paper of this book, we will not repeat it here.

3.2   Propagation, reflection and interaction of nonlinear equations

Since the end of seventies mathematicians began to study the topics on singularity analysis for nonlinear equations. J.M. Bony applied paradifferential operators to obtain a theorem on singularity propagation of solution to general nonlinear equation. J. Rauch, M. Beals and M. Reed, Chen Shuxing studied this topics by using pseudodifferential operators with nonsmooth symbols, and obtain a theorem on singularity propagation of solution for semilinear equations in [A2] and for general nonlinear hyperbolic equation in [CS1]. The authors of [C&L1] introduced the Fourier integral operators with finitely smooth phase and amplitude. The author of [QQ6] introduced the concept of para–Fourier integral operators and proved the Egorov's theorem for paradifferential operators. Then a theorem of singularity propagation for nonlinear operators of generalized principal type can be established, this result extended the corresponding conclusion in [A3].

Interactions of singularities are special phenomena for nonlinear equations. Such phenomena for hyperbolic system in one space–dimensional case were firstly analyzed by J. Rauch and M. Reed (see [A11]). Chen Shuxing studied such problem for semilinear hyperbolic equations

$$P_m(x, D) = f(x, u, ..., D^{m-1}u) \tag{3.4}$$

of higher order in multidimensional space. In [CS2], [CS6] he introduced the concept of admissible function and fused function on cotangent bundle, then one can define regularity index of solution according to the initial data and obtain

Theorem 3.3 Assume that u is an H^s solution of (3.4) satisfying the initial data $\partial^j u|_{x_n=0} = g_j (j=1,..., m-1)$, $s < \frac{n+1}{2} + m + 1$, $r(x, \xi)$ is the regularity index determined by the data g_j, then the regularity function $s_u(x, \xi)$ of the solution u satisfies $s_u(x, \xi) \geq r(x, \xi)$.

The regularity index can also be applied to discuss the reflection of singularities on boundary, (see [CS3]).

For some given nonlinear equations people may obtain better conclusion on singularity propagation than that for general equations due to their special form. By using the convexity of characteristic cone of strictly hyperbolic equation of second order, Liu Linqi in [LL1] proved the theorem of 3s–regularity propagation for the solutions of $p_2(x, D) = f(x, u, Du)$. However, such a theorem is not valid for equations of higher order (see [CS4]).

Xu Chaojiang also studied the problem on reflection of singularities for fully nonlinear equation of second order. On the basis of other works he considered the case when the bicharacteristic strip is tangent to the boundary in higher order. He proved that near the glancing set G (the definition is given in [A10]), the sigularities of solutions propagate along the generalized bicharacteristies.

The study on triple interaction has also been developed in China, Yu Yuenian in [YY1] concerned the Cauchy problem for semilinear wave equations, whose initial data have conormal singularities on finite curves intersecting at one point on the initial plane. The author proved that the solution is of conormal type, and its singularities are contained in the union of the characteristic surfaces through these curves and the characteristic cone issuing from the intersecting point. Comparing with [A3] and [A11], the smoothness loss of the solution is almost avoided. [CS13] is devoted to the triple interaction for nonlinear boundary value problems. The reflection and interaction of progressing waves are considered simultaneously.

Recently, microlocal analysis is also applied to discuss the problems on existence and smoothness of solutions with strong discontinuity for quasilinear hyperbolic system. [CS10] studied the solution with jump for hyperbolic system of conservation laws

$$\frac{\partial u}{\partial t} + \sum \frac{\partial f_i}{\partial x_i} = 0 \qquad (3.5)$$

On the surface S: $\phi(t, x) = 0$ bearing the jump of the solution u satisfies the conditions

$$[u]\phi_t + \sum [f_i]\phi_{x_i} = 0 \qquad (3.6)$$

where [] represents the jump of corresponding functions. Chen Shuxing discussed paradifferential operators with a parameter and their properties, and then by using them proved.

Theorem 3.4: Assume that u is a solution of the system (3.5), S is the surface bearing strong discontinuity, $u^{\pm}$ (the function u on both sides of S) belongs to H^s $(S \geq \frac{n+7}{2})$, on S the condition (3.6) holds. Moreover, assume that the genuine nonlinearity and the uniform Lopatinski condition are satisfied. Then $u^{\pm} \in H^{s_1}$ and $\phi \in H^{s_1+1}(s_1 > s)$ for $t > 0$, provided $u^{\pm} \in H^{s_1}$ and $\phi \in H^{s_1+1}$ hold for $t < 0$.

We refer readers to [CS11], [LD1] and [LD2], where the study on existence of solution to nonlinear hyperbolic system by means of microlocal analysis is developed. The results in these works have been mentioned in another paper of this book.

Reference

[CK1]. Chang Kungching, On the L^p continuity of the pseudo-differential operators, Scientia Sinica, Ser. A, 17(1974), 621-638.

[CS1]. Chen Shuxing, Pseudodifferential operators with finitely smooth symbols and their applications to quailinear equatins, Nonlinear Analysis, 6(1982), 1193-1206.

[CS2]. Chen Shuxing, Regularity estimate of solution to semilinear wave equation in higher space dimension, Scientia Sinica, Ser. A, 27(1984), 924-935.

[CS3]. Chen Shuxing, The reflection and interaction of the singularities of solutions to semilinear wave equation in higher space dimension, Nonlinear Analysis TMA, 8(1984), 1167-1179.

[CS4]. Chen Shuxing, Propagation of anomalus singularities of solutions to semilinear hyperbolic equation of higher order, Northeastern Math. Journal, 1(1985), 127-137.

[CS5]. Chen Shuxing, Regularity estimate of solutions to semilinear hyperbolic equations in higher space dimension, Acta MAth. Sinica, 3(1987), 66-76.

[CS6]. Chen Shuxing, Global multi-Holder estimate of solutions to elliptic equations of higher order, Chin. Ann. Math., 8A(1987), 239-251.

[CS7]. Chen Shuxing, On the propagation of singularities of the solutions for nonlinear systems, Proceedings of the Changchun Symposium on P.D.E., 1986, 309-316.

[CS8]. Chen Shuxing, Necessary conditions on local solvability of boundary value problems for linear PDEs, Kexue Tongbao, 33(1988), 617-621.

[CS9]. Chen Shuxing, Smoothness of shock front solutions for system of conservation laws, Lecture Notes in Math., Springer-Verlag, # 1306 (1988), 38-60.

[CS10]. Chen Shuxing, On reflection of multidimensional shock front, Jour. Diff. Eqs., 80(1989), 199-236.

[CS11]. Chen Shuxing, Reflection and interaction of progressing wave for semilinear wave equations, Jour. Math. Anal. Appl., 153(1990), 562-575.

[CS12]. Chen Shuxing, Piecewise smooth solutions of semilinear systems in higher space dimension, Chin. Math. Ann., 10B(1989), 361-370.

[C&L1]. Chen Shuxing & Liu Linqi, Fourier integral operators of finite grade and their applications, (I), (II), Chin. Ann. Math., 6A(1985), 83-94, 323-334.

[CZ1]. Cui Zhiyong, Complex Maslov line bundles. Northeastern Math. J., 2(1986), 120-126.

[DX1]. Du Xinhua, A necessary condition of local solvability for a class of non-pricipal type, Acta Math. Sinica, 29(1986), 490-493.

[HJ1]. Hong Jiaxing, On Microlocal analysis for a class of operqtors with multiple characteristics, Acta Math. Sinica, 28(1985), 23-34.

[HJ2]. Hong Jiaxing, The theorem of partial hypoellipticity in characteristic case and its applications, Acta Math. Sinica, 29(1986), 327-337.

[HJ3]. Hong Jiaxing, Singular direction of WF(u) and its application, Chin. Ann. Math., 7B(1986).

[HJ4]. Hong Jiaxing, Reflection of singularities on the boundary, Chin. Ann. Math., 8B(1987).

[HY1]. Huang Yumin, On Egorov's principal type condition, Kexue Tongbao, 30(1985), 333-335.

[HY2]. Huang Yumin, Local estimates of pseudodifferential operators of non-principal type, Scientia Sinica, Ser. A, 28(1985), 801-813.

[HY3]. Huang Yumin, A class of partial differential operators of second order of non-principal type, Acta Math. Sinica.

[HY4]. Huang Yumin, On interior regularity of solutions of a class of hypoelliptic differential equations, Lecture Notes in Math., Springer-Verlag # 1306(1988), 93-101.

[HY5]. Huang Yumin, Local estimates of non-principal pseudodifferential operators, Scientia Sinica, 28(1985), 413-423.

[LC1]. Li Chenzhang, Remarks on the L^p-boundedness of pseudodifferential operators, Proc. 1982 Changchun symp., Science Press, Beijing, (1986), 463-472.

[LC2]. Li Chenzhang, On the boundedness of pseudodifferential operators in space $H_p^{s/a}(\mathbb{R}_+^n)$, Northeastern Math. J., 2(1986), 186-195.

[LD1]. Li Dening, Stability of shock waves for multi-dimensional hyperbolic-parabolic conservation laws, Scientia Sinica, Ser. A, 31(1988), 1-15.

[LD2]. Li Dening, The nonlinear initial-boundary value problem and the existence of multi-dimensional shock wave quasilinear hyperbolic-parabolic coupled systems, Chin. Ann. Math., 8B(1987), 252-280.

[LLq1]. Liu Linqi, Optimal propagation of singularities for semilinear hyperbolic differential equations, Ph. D Thesis, Inst. Math. Fudan Univ. (1985).

[LLj1]. Lu Lijiang, Cauchy problem of a class of of pseudo-differential operators, Acta Math. Sinica, 26(1983), 114-128.

[LZ1]. Lu Zhujia, Discrete phenomena in existence in Goursat problem for equation with double characteristics, Scientia Sinica, Ser. A, 26(1983), 595-606.

[LZ2]. Lu Zhujia, Mai Mingcheng, Wang Guangyin, Discrete phenomena in existence in the initial value problems, Scientia Sinica, Ser. A, 22(1979), 1229-1237.

[LW1]. Luan Wengui, Uniqueness and bounds of solution of Cauchy problem for first order pseudodifferential equations, Scientia Sinica, Ser. A, 26(1983), 225-238.

[LX1]. Luan Xuebo, On multiplication in $\mathcal{F}$, solvability and hypoellipticity of a class of LPDO in $\mathcal{F}$, Kexue Tongbao, 29(1984), 1416-1418.

[L& F1]. Luo Xuebo and Fu Chuli, A class of hypoelliptic LPDO not of principal type, Acta Math. Sinica, 28(1985), 233-243.

M & L1]. Mai Mingcheng and Lu Zhujia, On the uniqueness in Cauchy problems for linear partial differential equations, Acta Math. Sinica, 22(1979), 713-718.

[QM1]. Qi Mingyou, On partial differential equations with a manifold of regular singularity of lower dimension, Kexue Tongbao, 28(1983), 740-743.

[QQ1]. Qiu Qingjiu, Parametrices and local solvability of the pseudodifferential operators $t^m \partial_t + B(x, t, D_x)$, Chin. Ann. Math., 2B(1981), 59-64.

[QQ2]. Qiu Qingjiu, Local solvability and propagation of singularities for a class of pseudodifferential operators with irregular singularity, Chin. Ann. Math., 3A(1982), 57-66.

[QQ3]. Qiu Qingjiu, The Besov space boundedness for certain Fourier integral operators, Acta Math. Sinica, 5(1985), 167-174.

[QQ4]. Qiu Qingjiu, L^p boundedness of Fourier operators with $S_{\rho,\delta}^m$ amplitude function, Scientia Sinica, Ser. A, 30(1987), 337-347.

[QQ5]. Qiu Qingjiu, On L^p-estimates for certain Fourier integral operators, Scientia Sinica, Ser. A, 31(1988), 350-362.

[QQ6]. Qiu Qingjiu, Para-Fourier integral operators, Scientia Sinica, Ser. A, 32(1989), 1036-1046.

[QQ7]. Qiu Qingjiu, On Egorov theorem for paradifferential operators, Scientia Sinica, Ser. A, 33(1990), 663-673.

[QQ8]. Qiu Qingjiu, Application of para-fourier integral operators to propagation of nonlinear singularities, Scientia Sinica, Ser. A, 33(1990), 1060-1071.

Q & Q1]. Qiu Qingjiu and Qian Sixin, Analysis of C^∞ singularities for a class of operators with varying multiple characteristics, Lecture Notes in Math., Springer-Verlag, # 1306 (1988), 141-148.

S& W1]. A. El Soufi and X.P. Wang, Some remarks on Witten's method, Poincare-Hopf theorem and Atiyah-Bott formula, Ann. Global Anal. Geom., 5(1987), 161-178.

[WC1]. Wang Chuanfang, On discrete phenomena in the mixed problem, Chin. Ann. Math., 1(1980), 469-476.

[WF1]. Wu Fangtong, On the fundamental solution of the Cauchy problem for a class of hyperbolic operators with principal part of Jordan form, Proc. 1982 Changchun Symp., Science Press, Beijing, (1986), 637-648.

[WF2]. Wu Fangtong, A class diffractive boundary value problem with multiple characteristics, Lecture Notes in Math., Springer-Verlag, # 1306(1988), 224-239.

[WML1]. Wang Guangyin, Mai Mingcheng and Lu Zhujia, On discrete phenomena of an initial problem, Kexue Tongbao, 23(1978), 279-282.

[WML2]. Wang Guangyin, Mai Mingcheng and Lu Zhujia, A remark about uniqueness for the partial differential equations of non-principal type, Acta Math. Sinica, 22(1979), 713-718.

[W& C1]. Wang Rouhuai and Cui Zhiyong, Generalized Leray formula on positive complex Lagrange-Grassmann manifolds, Chin. Ann. Math., 5B(1984), 215-234.

[W& L1]. Wang Rouhuai and Li Chengzhang, On the L^p boundedness of several classes of pseudo-differential operators, Chin. Ann. Math., 5B(1984), 193-213.

[WX1]. Wang Xueping, Comportement semiclassique de traces partielles, C.R.Acad. Sci. Paris, 299(1984), 867-870.

[WX2]. Wang Xueping, Asymptotic behaviour of spectral means for DO's, J. Appr. Theory and ite Appl., 1(1985), 119-136.

[WX3]. Wang Xueping, Puits multiple pour l'operateur de Dirac, Ann. Inst. Henry Poincare, 43A(1985), 269-319.

[WX4]. Wang Xueping, Etude semiclassique d'observables quantiques, Ann. Fac. Sci. Toulouse, 7(1985), 101-135.

[WX5]. Wang Xueping, Operateur de temps-retard dans la theorie de la diffusion, C.R.Acad. Sci. Paris, 301(1985), 789-792.

[WX6]. Wang Xueping, Approximation Semi-classique de l'equation de Heisenberg, Comm. in Math. Phys., 104(1986), 77-86.

[WX7]. Wang Xueping, Continuity of time-delay operators and low energy resolvent estimates, Proc. of Royal Soc. Edinburger, 105A(1987), 229-242.

[WX8]. Wang Xueping, Time-delay of scattering solutions and classical trajectories, Ann. Inst. Henri Poincare, 47(1987), 25-37.

[WX9]. Wang Xueping, Phase-space description of time-delay in scattering theory, Comm. in PDE, 13(1988), 223-259.

[WX10]. Wang Xueping, Time-delay of scattering solutions and resolvent estimates for semiclassical Schrödinger operators, J. Diff. Eqs., 71(1988), 348-395.

[XC1]. Xu Chaojiang, Regularite des solutions des d'equations derivees partielles non lineaires, C.R.Acad. Sci. 300(1985), 267-270.

[XC2]. Xu Chaojiang, Hypoellipticite d'equations aux derivees partielles non lineaires, Journees EDP, Saint-Jean-de-Mont, (1985).

[XC3]. Xu Chaojiang, Operateaus sous-elliptiques et regularites des solutions des equations aux derivees partielles non lineaires du second order dans R, Comm. PDE, 11(1986), 1575-1603.

[XC4]. Xu Chaojiang, Regularite des solutions d'equations aux derivees partielles non lineaires associees a un systeme de champs de vecteurs, Ann. Inst. Fourier, 37(1987), 105-113.

[XC5]. Xu Chaojiang, Hypoellipticity of nonlinear second order partial differential equations, Journal PDE, 1(1988), 85-95.

[XC6]. Xu Chaojiang, On the regularity of the minima of quasi-convex integrals (preprint).

[XC7]. Xu Chaojiang, Propagation au bord des singularites pour des problemes de dirichlet non lineaires d'ordre deux (preprint).

X$ Z1]. C.J. Xu and C. Zuily, Smoothness up to boundary for solutions of nonlinear and nonelliptic Dirichlet problem, Trans. Amer. Math. Soc., 306(1988), 1-15.

[YY1]. Yu Yuenian, Sigularities of solutions to Cauchy problems for semilinear wave equations in two space dimensions, Jr. PDEs, 3(1990), 69-80.

[YY2]. Yu Yuenian, Piecewise smooth solution to quasilinear hyperbolic systems, Chin. Math. Ann., 11A(1990), 104-110.

[ZX1]. Zheng Xingli, On initial value problem of a class of degenerate equation, Kexue Tongbao, 28(1983), 1433-1436.

[ZX2]. Zheng Xingli, Existence of solutions to Cauchy problem of an equation with double characteristics, Scientia Sinica, Ser. A, 28(1985), 357-367.

Additional References

[A1]. M. Beals, L Boundedness of Fourier integral operators, Memoirs of Amer. Math. Soc., 264(1982).

[A2]. M. Beals & M. Reed, Propagation of singularities for hyperbolic pseudodifferential operators with non-smooth coefficients, Comm. Pure Appl. Math., 35(1982), 169-184.

[A3]. J.M. Bony, Calcul symbolique et propagation des singularites pour les equations aux derivees partielles non lineaires. Ann. Scien. Ecole Norm. Sup., 4 Serie, t.14 (1981), 209-246.

[A4]. R. Coifman et Y. Meyer, Au dela des operateurs pseudo-differentiels, Asterisque, 57(1978).

[A5]. C. Fefferman and D. Phong, The uncertainty principle and sharp Garding inequalities, Comm. Pure Appl. Math., 34(1981), 285-331.

[A6]. L. Hörmander, Pseudo-differential operators and hypoelliptic equations, Amer. Math. Soc. Symp. on Singular Integrals, (1966), 138-183.

[A7]. L. Hörmander, Hypoelliptic second order differential equations. Acta Math. 119(1967), 147-171.

[A8]. L. Hörmander, The analysis of linear differential operators, Springer-Verlag, Berlin Heidelberg New York Tokyo, 1985.

[A9]. R.B. Melrose, Microlocal parametrices for diffractive boundary value problems, Duke Math. J., 42(1975), 605-635.

[A10]. R.B. Melrose and J. Sjostrand, Singularities of boundary value problems I, II, Comm. Pure Appl. Math., 31(1978), 593-617; 35(1982), 129-168.

[A11]. R.B. Melrose and N. Ritter, Interaction of progressing waves, Annals of Math., 121(1985), 187-213.

[A12]. J. Rauch and M. Reed, Nonlinear microlocal analysis of semilinear hyperbolic systems in one space dimension, Duke Math. J., 49(1982), 397-475.

[A13]. L. Rothschild and E. Stein, Hypoelliptic differential operators and nilpotent groups, Acta Math., 137(1976), 247-320.

[A14]. F. Treves, Discrete phenomena in the Cauchy problems. Proc. Amer. Math. Soc., 46(1974), 229-233.

Nonlinear Partial Differential Equations in Physics and Mechanics†

Yulin Zhou and Boling Guo

Centre for Nonlinear Studies Institute of Applied Physics and
Computational Mathematics Beijing, P.O.Box 8009

1   The Landau-Lifshitz Equation of the Ferromagnetic Chain

Let Ω be a domain in m-dimensional Enclidean space R^m. The n-dimensional classical system for the isotropic Heisenberg chain with spin density $Z_i(x,t)$ $(i = 1, 2, 3)$ is described by the Hamiltonian density

$$H = \frac{\alpha_2}{2}|\nabla \vec{Z}|^2 - H_0\vec{Z}$$

where α_2 is the exchange constant, $\vec{Z}$ is the spin vector and H_0 an external magnetic field. The spin equation of motion with Gilbert damping term (without the external magnetic field) has the form

$$\partial_t \vec{Z} = -\alpha_1 \vec{Z} \times (\vec{Z} \times \triangle \vec{Z}) + \alpha_2 \vec{Z} \times \triangle \vec{Z}, \tag{0.1}$$

where "$\times$" denotes the vector cross product in R^3, $\vec{Z} = (Z_1, Z_2, Z_3) : \Omega \times [0, T) \to R^3$ is the spin vector and $\alpha_1 \geq 0$ is a Gilbert damping constant (see [1]). The system (0.1) is implied by the conservation of energy and magnitude of $\vec{Z}$, and is a version which gives rise to a continuum spin wave theory.

The continuous Heisenberg spin chain has aroused considerable interest among physicists. The above processional equation (0.1) of motion was first derived on phenomenological grounds by Landau-Lifshitz [2]. The Equation (0.1) bears on a fundamental role in the understanding of nonequilibrium magnetism. A lot of work contributed to the study of the soliton for the Landau-Lifshitz equation of the 1-dimensional motion spin chain has been made by physicists and mathematicians [3][4][5] [6].

---

†1991 Mathematics Subject Classification: 35Q
† Supported by the National Natural Science Fund of China

127

C. Gu et al. (eds.), Partial Differential Equations in China, 127–159.
© 1994 *Kluwer Academic Publishers*.

From the mathematical views, the equation (0.1) is interesting when $\alpha_1 = 0$, we can write the equation (0.1) as the following form:

$$\overrightarrow{Z}_t = A(\overrightarrow{Z})\triangle\overrightarrow{Z}, \qquad (0.2)$$

where

$$\overrightarrow{Z} = \begin{pmatrix} Z_1 \\ Z_2 \\ Z_3 \end{pmatrix}, \quad A(\overrightarrow{Z}) = \begin{pmatrix} 0 & -Z_3 & Z_2 \\ Z_3 & 0 & -Z_1 \\ -Z_2 & Z_1 & 0 \end{pmatrix}. \qquad (0.3)$$

It is not difficult to verify that:

(i) the matrix $A(\overrightarrow{Z})$ is "zero definite", i.e.,

$$\overrightarrow{\xi} \cdot A(\overrightarrow{Z})\,\overrightarrow{\xi} = 0, \quad \forall\, \overrightarrow{\xi},\, \overrightarrow{Z} \in R^3.$$

(ii) $A(\overrightarrow{Z})$ is singular, i.e.,

$$\det A(\overrightarrow{Z}) = 0, \quad \forall\, \overrightarrow{Z} \in R^3.$$

In fact, $A(\overrightarrow{Z})$ is a three order symplectic matrix. So the equation (0.2) is a strongly degenerate and strongly coupled quasiparabolic system. It is difficult to deal with this problem by the ordinary way of parabolic partial differential equations.

On the other hand, we find some new links between harmonic maps and the solution of the Landau-Lifshitz equation of the ferromagnetic spin chain. More precisely, the elliptic type Landau-Lifshitz equation, i.e. $\partial_t \overrightarrow{Z} = 0$ in (0.1) (Laplacian $\triangle$ replaced by Laplace- Betrami operator $\triangle_M$), is equivalent to the harmonic map equation $M \to S^2$. Moreover, when $\alpha_2 = 0$ in (0.1), it can be proven that the Landau-Lifshitz equation has the same form as the of the heat flow for harmonic maps.

Since 1982, we have studied systematically the various problems for Landau-Lifshitz equation (0.2) and (0.1) by using the different proving flame. For example, in the case of one dimension we apply the viscous elimination method and Leray-Schauder argument to solve the Cauchy problem and the initial-boundary value problem, and the nonlinear boundary value problem by using the finite difference-differential method. For m ($m > 1$) dimensions, it is to be found that it is difficult to solve this problem by the mentioned method, we apply the Galerkin method to get the generalized global solution for the initial-boundary value problem of Landau-Lifshitz equation in m dimensions [see 7-11]. We also consider the Landau-Lifshitz equation in Compact Riemannian manifolds [12] and geometrical extensions for systems of ferromagnetic chain [13].

In this section we state the main results for the Landau-Lifshitz equation.

1.1 One dimensional Cases

First, we consider the following initial-boundary value problems for the generalized Landau-Lifshitz equation

$$\overrightarrow{Z}_t = \overrightarrow{Z} \times \overrightarrow{Z}_{xx} + \overrightarrow{f}(x,t,\overrightarrow{Z}), \quad t > 0,\ 0 < x < l \qquad (1.1)$$

$$\vec{Z}\big|_{t=0} = \vec{Z}_0(x), \quad 0 \le x \le l \tag{1.2}$$

with the first boundary condition

$$\vec{Z}(0,t) = \vec{Z}(l,t) = 0, \tag{1.3}$$

or the second boundary condition

$$\vec{Z}_x(0,t) = \vec{Z}_x(l,t) = 0, \tag{1.4}$$

or the mixed boundary condition

$$\vec{Z}(0,t) = \vec{Z}_x(l,t) = 0, \tag{1.5}$$

or

$$\vec{Z}_x(0,t) = \vec{Z}(l,t) = 0, \tag{1.6}$$

where $\vec{Z}(x,t) = (Z_1(x,t), Z_2(x,t), Z_3(x,t))$ is the 3-dimensional unknown vector function, "$\times$" denotes the cross product between two vectors $\vec{\xi} \in R^3$. $\vec{f}(x,t,\vec{Z})$ is the 3- dimensional vector function of variables $x \in R$, $t \in R^+$, $\vec{Z} \in R^3$.

We apply the Leray-Schauder fixed point argument to prove the existence of global weak solution of the initial-boundary value problems.

Let us consider the linear parabolic system

$$u_t - A(x,t)u_{xx} + B(x,t)u_x + C(x,t)u = f(x,t) \tag{1.7}$$

with the following initial and boundary conditions

$$u(0,t) = 0 \text{ or } u_x(0,t) = 0, \text{ at } x = 0 \tag{1.8}$$

$$u(l,t) = 0 \text{ or } u_x(l,t) = 0, \text{ at } x = l \tag{1.9}$$

$$u(x,0) = u_0(x), \text{ at } t = 0 \tag{1.10}$$

Lemma 1.1 Suppose that the linear parabolic system (1.7) satisfies the following conditions
(1) $A(x,t)$ is a $N \times N$ positively definite matrix in $Q_T = [0,l] \times [0,T]$,
(2) $A(x,t), B(x,t), C(x,t)$ are $N \times N$ mensurable and bounded matrixes.
(3) $f(x,t) \in L_2(Q_T)$,
(4) $u_0(x) \in W_2^{(1)}(0,l)$ satisfying the boundary conditions.
Then there exists a unique vector valued solution $u(x,t) \in L_\infty(0,T; W_2^{(1)}(0,l)) \cap W_2^{(2,1)}(Q_T)$ and holds the following estimate

$$\sup_{0 \le t \le T} \|u(\cdot,t)\|_{W_2^{(1)}(0,l)} + \|u_t\|_{L_2(Q_T)} + \|u_{xx}\|_{L_2(Q_T)}$$

$$\le K(\|u_0\|_{W_2^{(1)}(0,l)} + \|f\|_{L_2(Q_T)}). \tag{1.11}$$

Consider the spin system

$$\vec{Z}_t = \epsilon \vec{Z}_{xx} + \vec{Z} \times \vec{Z}_{xx} + \vec{f}(x,t,\vec{Z}), \quad \epsilon > 0 \tag{1.12}$$

with the same boundary value conditions.

Take the functional space $B = L_\infty(Q_T)$ as the base space for fixed point argument. We define the functional mapping: $T_\lambda : B \to B$ of the base space into itself with parameter $0 \le \lambda \le 1$ as follows:

For every $\vec{u} \in B$, the image $Z = T_\lambda(\vec{u})$ is the solution of the initial-boundary value problem of the linear parabolic system

$$\vec{Z}_t = \epsilon \vec{Z}_{xx} + \lambda \vec{u} \times \vec{Z}_{xx} + \lambda \vec{f}(x, t, \vec{u}).$$

By Lemma 1, $\vec{Z} = T_\lambda \vec{u}$ is uniquely determined and belongs to the space

$$\vec{Z} \in L_\infty(0, T; W_2^{(1)}(0, l)) \cap W_2^{(2,1)}(Q_T).$$

It is clear that the functional operator T_λ for every λ is completely continuous for any bounded set M of the base space B, the operator T_λ is uniformly continous with respect to $\lambda \in [0, 1]$.

In order to prove the existence of solution for the generalized spin system by means of the Leray-Schauder fixed point theorem, we shall establish a priori uniform estimates for all possible fixed points of the mapping T_λ with respect to the paramater $\lambda \in [0, 1]$.

For this purpose, we make the following assumptions:

(I) $\vec{f}(x, t, \vec{Z})$ is continuously differentiable with respect to x and $\vec{Z}$. The 3×3 Jacobi derivative matrix $\dfrac{\partial \vec{f}}{\partial \vec{Z}}(x, t, \vec{Z})$ is semi-bounded with a constant, i.e. for any 3-dimensional vector $\xi \in R^3$, the inequality

$$\vec{\xi} \cdot \vec{f}_Z(x, t, \vec{Z}) \vec{\xi} \le b|\vec{\xi}|^2 \tag{1.13}$$

holds, and $\vec{f}_0(x, t) = \vec{f}(x, t, 0) \in L_2(Q_T)$.

(II) For $(x, t, \vec{Z}) \in Q_T \times R^3$, there is the inequality

$$|\vec{f}_x(x, t, \vec{Z})| \le c(x, t)|\vec{Z}|^3 + d(x, t) \tag{1.14}$$

where $c(x, t) \in L_\infty(Q_T)$, and $d(x, t) \in L_2(Q_T)$.

(III) $\vec{Z}_0(x) \in W_2^{(1)}(0, l)$, and satisfies the boundary conditions.

By means of the integral estimates, we have

$$\sup_{0 \le t \le T} \|\vec{Z}(\cdot, t)\|_{L_2(0, l)} + \sup_{0 \le t \le T} \|\vec{Z}_x(\cdot, t)\|_{L_2(0, l)} \le K_1, \tag{1.15}$$

where K_1 is a constant independent of ϵ, λ and l.

Theorem 1.1 Under the conditions (I)-(III), the second boundary value problem of the nonlinear parabolic system (1.12) with $\epsilon > 0$ has a generalized global solution

$$\vec{Z}(x, t) \in L_\infty(0, T; W_2^{(1)}(0, l)) \cap W_2^{(2,1)}(Q_T).$$

Theorem 1.2 Suppose that the conditions (I)-(III) are satisfied, and the spin system is homogeneous, i.e. $\overrightarrow{f}(x,t,0) = 0$. Then the first boundary problem and the mixed boundary problem for the spin system (1.12) with $\epsilon > 0$ has a generalized global solution

$$\overrightarrow{Z} \in L_\infty(0,T;W_2^{(1)}(0,l) \cap W_2^{(2,1)}(Q_T).$$

Theorem 1.3 Under the conditions of Theorem 1.1 and Theorem 1.2 in the domain $Q_\infty = \{0 \leq x \leq l, t \in R^+\}$, if $b < 0$, then we have

$$\lim_{t \to \infty} \|\overrightarrow{Z}(\cdot,t)\|_{L_2(0,l)} = 0.$$

Now we consider the initial-boundary problem for the system of ferromagnetic chain ($\epsilon = 0$)

$$\overrightarrow{Z}_t = \overrightarrow{Z} \times \overrightarrow{Z}_{xx} + \overrightarrow{f}(x,t,\overrightarrow{Z}).$$

One can establish the a priori estimates for the spin system and which are independent of ϵ

$$\|\overrightarrow{Z}_\epsilon(\cdot,t)\|_{L_2(0,l)} \leq K_3(\|\overrightarrow{Z}_0\|_{L_2(0,l)} + \|\overrightarrow{f}_0\|_{L_2(0,l)})e^{(b+\delta)t},$$

$$\sup_{0 \leq t \leq T} \|\overrightarrow{Z}_\epsilon(\cdot,t)\|_{W_2^{(1)}(0,l)} \leq K_4,$$

$$\sup_{0 \leq t \leq T} \|\overrightarrow{Z}_{\epsilon t}(\cdot,t)\|_{H^{-1}(0,l)} \leq K_5,$$

$$\|\overrightarrow{Z}_\epsilon(x,t)\|_{C^{(1/2,1/4)}(Q_T)} \leq (1+l)K_6, \tag{1.16}$$

where the constants K_3, K_4, K_5 and K_6 are independent of ϵ.

Let $\epsilon \to 0$, we can get

Theorem 1.4 Suppose that the system of ferromagnetic chain and the initial vector function $\overrightarrow{Z}_0(x)$ satisfy the conditions (I)-(III). The second boundary problem of the ferromagnetic chain has as least one global weak solution

$$\overrightarrow{Z}(x,t) \in L_\infty(0,T;W_2^{(1)}(0,l)) \cap C^{(1/2,1/4)}(Q_T).$$

Theorem 1.5 Suppose that the conditions of Theorem 1.4 are satisfied, and $\overrightarrow{f}(x,t,0) = 0$. Then the first boundary problem or the mixed boundary problem for the system of ferromagnetic chain has as least one global weak solution

$$\overrightarrow{Z}(x,t) \in L_\infty(0,T;W_2^{(1)}(0,l)) \cap C^{(1/2,1/4)}(Q_T).$$

Theorem 1.6 If $b < 0$, then we have

$$\lim_{t \to \infty} \|\overrightarrow{Z}(\cdot,t)\|_{L_2(0,l)} = 0. \tag{1.17}$$

For the first boundary problem

$$\overrightarrow{Z}(0,t) = 0, \ \overrightarrow{Z}(x,0) = \varphi(x), \ 0 < x < \infty \tag{1.18}$$

and the second boundary problem

$$\vec{Z}_x(0,t) = 0, \ \vec{Z}(x,0) = \varphi(x), \ 0 < x < \infty \tag{1.19}$$

in the semi-infinite domain $Q_T^* = \{x \in R^+, 0 \le t \le T\}$, we have

Theorem 1.7 Suppose that $\vec{f}(x,t,\vec{Z})$ and $\vec{Z}_0(x)$ satisfy the conditions (I)-(III) in Q_T^* and also satisfy either of the following conditions

(IV$_1^*$) For $(x,t,\vec{Z}) \in Q_T^* \times R^3$

$$|\vec{f}(x,t,\vec{Z})| \le \bar{c}(x,t)F(\vec{Z}) + \bar{d}(x,t), \tag{1.20}$$

where $F(\vec{Z})$ is a continuous function of $\vec{Z} \in R^3$, and $\bar{c}(x,t)$ and $\bar{d}(x,t) \in L_\infty(0,T;L_2(R^+))$. or

(IV$_2^*$) For $(x,t,\vec{Z}) \in Q_T^* \times R^3$

$$|\vec{f}(x,t,\vec{Z})| \le c(x,t)|\vec{Z}|^l + \bar{d}(x,t), \tag{1.21}$$

where $l \ge 0$, $c(x,t) \in L_\infty(Q_T^*)$, $\bar{d}(x,t) \in L_\infty(0,T;L_s(R^+))$, $2 \ge s > 1$.
Then for the second boundary problem for the system of ferromagnetic chain has at least one global weak solution

$$\vec{Z}(x,t) \in L_\infty(0,T;W_2^{(1)}(R^+)) \cap C_{loc}^{(1/2,1/4)}(Q_T^*).$$

Now we are going to consider the Cauchy problem associated with the system of ferromagnetic chain with the Gilbert damping term

$$\vec{Z}_t = -\epsilon \vec{Z} \times (\vec{Z} \times \vec{Z}_{xx}) + \vec{Z} \times \vec{Z}_{xx}, \ (\epsilon \ge 0) \tag{1.22}$$

$$\vec{Z}|_{t=0} = \vec{Z}_0(x), \ |\vec{Z}_0(x)| = 1. \tag{1.23}$$

We are first concerned with the following diffusion problem:

$$\vec{Z}_t = \epsilon \vec{Z}_{xx} + \vec{Z} \times \vec{Z}_{xx} + \epsilon |\vec{Z}_x|^2 \vec{Z}, \ \epsilon > 0 \tag{1.24}$$

$$\vec{Z}(x,0) = \vec{Z}_0(x), \tag{1.25}$$

$$\vec{Z}(x+D,t) = \vec{Z}(x-D,t), \tag{1.26}$$

where D is a positive number. The existence of smooth solution to the problem (1.24)-(1.26) is verified by the spatial difference method and a priori estimates of higher-order derivatives in Sobolev spaces. In the same time we can show that the initial value problem with the periodic condition (1.26) for equation (1.24) with $\epsilon > 0$ is equivalent in the classical sense to the problem for system of ferromagnetic chain with the Gilbert damping term, i.e.

$$\vec{Z}_t = -\epsilon \vec{Z} \times (\vec{Z} \times \vec{Z}_{xx}) + \vec{Z} \times \vec{Z}_{xx}. \tag{1.27}$$

Then we establish uniform estimates for the smooth solution of the problem (1.24)-(1.26) relative to ϵ and D. We pass to the limit as $\epsilon \to 0$ and $D \to \infty$, and achieve to get the existence of a unique global smooth solution of problem (1.27)(1.25)(1.26) with $\epsilon = 0$, and solution of the Cauchy problem (1.27)(1.25) with $\epsilon = 0$.

Theorem 1.8[14] Let ϵ be any positive number and suppose that $\overrightarrow{Z}_0(x) \in H^k(\Omega)$ and $\overrightarrow{Z}_0$ satisfies: $|\overrightarrow{Z}_0(x)| = 1$, $\overrightarrow{Z}_0(x - D) = \overrightarrow{Z}_0(x + D)$ for any $x \in \Omega$ with $D > 0$. Then the initial value problem with the periodic boundary conditions (1.25)(1.26) for the system of ferromagnetic chain (1.27) has at least one global smooth solution $\overrightarrow{Z}(x,t) \in G = \sum_{s=0}^{[k/2]} W_\infty^s(0,T; H^{k-2s}) \cap \cap_{s=0}^{[(k+1)/2)]} H^s(0,T; H^{k+1-2s})$, where s, k are nonnegative integers and T is an arbitrary positive number.

Lemma 1.2 Under the conditions of Theorem 1.8, let $\overrightarrow{Z} = \overrightarrow{Z}(x,t)$ be a smooth solution of the problem (1.24)-(1.26) with $\epsilon > 0$. Then we have

$$\sup_{0 \le t \le T} \|\overrightarrow{Z}_{x^{k-2s}t^s}(\cdot, t)\|_{L_2(\Omega)} \le C, \tag{1.28}$$

where C is a constant independent of D and ϵ, k, s are nonnegative integers with $k - 2s \ge 0$.

Upon employing the uniform estimate (1.28) and a standard method, one could prove the following existence result of the problem (1.27) (1.25) (1.26) with $\epsilon = 0$ by passing to the limit in equation (1.24) or (1.27) as $\epsilon \to 0$.

Theorem 1.9 Let the initial data $\overrightarrow{Z}_0(x)$ be given in $H^k(\Omega)$, such that $|\overrightarrow{Z}_0(x)| = 1$, $\overrightarrow{Z}_0(x - D) = \overrightarrow{Z}_0(x + D)$ for any $x \in R^1$, here D is a positive number. Then for the initial value problem with periodic boundary condition of the degenerated system of ferromagnetic chain (1.27) with $\epsilon = 0$, there exists a unique global smooth solution $\overrightarrow{Z}(x,t)$ such that

$$\overrightarrow{Z}(x,t) \in \cap_{s=0}^{[k/2]} W_\infty^s(0,T; H^{k-2s}(\Omega)), \quad k \ge 4.$$

Theorem 1.10 Let ϵ be an arbitrary nonnegative number. Let the initial data $\overrightarrow{Z}_0(x)$ be given in $H^{k-1}(R^1)$ with $k \ge 4$ such that $|\overrightarrow{Z}_0(x)| = 1$, for $x \in R^1$. Then the Cauchy problem for the system of ferromagnetic chain (1.27) has a unique global smooth solution $\overrightarrow{Z}(x,t)$ such that

$$\overrightarrow{Z}(x,t) \in \cap_{s=0}^{[k/2]} W_\infty^s(0,T; H^{k-2s}(R^1)),$$

with $k - 2s \ge 0$.

Finally, we consider some nonlinear boundary value problem for the system

$$\overrightarrow{Z}_t = \overrightarrow{Z} \times \overrightarrow{Z}_{xx} + \overrightarrow{f}(x,t,\overrightarrow{Z}), \tag{1.29}$$

and the corresponding system

$$\overrightarrow{Z}_t = \epsilon \overrightarrow{Z}_{xx} + \overrightarrow{Z} \times \overrightarrow{Z}_{xx} + \overrightarrow{f}(x,t,\overrightarrow{Z}). \tag{1.30}$$

Let us give the nonlinear boundary conditions

$$\vec{Z}_x(0,t) = \text{grad}\psi_0(t, \vec{Z}(0,t)),$$

$$-\vec{Z}_x(l,t) = \text{grad}\psi_1(t, \vec{Z}(l,t)), \tag{1.31}$$

and the initial condition

$$\vec{Z}(x,0) = \vec{\varphi}(x), \tag{1.32}$$

where $\psi_0(t, \vec{Z})$ and $\psi_1(t, \vec{Z})$ are scalar functions, "grad" denotes the gradient operator with respect to $\vec{Z}$. This class of nonlinear boundary problems appear in many practical problems.

We also consider the mixed boundary problem with condition

$$\vec{Z}_x(0,t) = \text{grad}\psi_0(t, \vec{Z}(0,t)),$$

$$\vec{Z}(l,t) = 0, \tag{1.33}$$

and the boundary problem in the semi-infinite domain $Q_T^* = \{x \in R^+, 0 \le t \le T\}$

$$\vec{Z}_x(0,t) = \text{grad}\psi_0(t, \vec{Z}(0,t)), \ t > 0, 0 < x < \infty$$

$$\vec{Z}(x,0) = \varphi(x), \ 0 \le x < \infty. \tag{1.34}$$

In the cases of nonlinear boundary problems, it is difficult to prove the existence of global weak solution by using Leray-Schauder fixed point method and Galerkin method. Here we apply the finite difference- differential method. From the practice, this shows that it is successful and useful.

The finite interval $[0, l]$ is divided into the small segment grids by the point $x_j = jh \ (j = 0, 1, \cdots, J)$, where $Jh = l$. The function $Z_h(t) = \{Z_j(t)|_{j=0,\cdots,J}\}$ is defined on grid points $x_j = jh \ (j = 0, 1, \cdots, J)$.

Construct the system of ordinary differential equations

$$\vec{Z}'_j(t) = \epsilon\frac{\Delta_+\Delta_-\vec{Z}_j(t)}{h^2} + \vec{Z}_j(t) \times \frac{\Delta_+\Delta_-\vec{Z}_j(t)}{h^2} + \vec{f}(x, t, \vec{Z}_j(t)) \tag{1.35}$$

with nonlinear boundary conditions

$$\frac{\Delta_+\vec{Z}_0(t)}{h} = \text{grad}\psi_0(t, \vec{Z}_0(t)),$$

$$-\frac{\Delta_-\vec{Z}_J(t)}{h} = \text{grad}\psi_1(t, \vec{Z}_J(t)), \tag{1.36}$$

and the initial conditions

$$\vec{Z}_j(0) = \vec{\varphi}_j, \ j = 0, 1, \cdots, J \tag{1.37}$$

where $\vec{\varphi}_j = \vec{\varphi}(x_j)$ $(j = 1, \cdots, J - 1)$, and $\vec{\varphi}_0$ and $\vec{\varphi}_J$ satisfying

$$\vec{\varphi}_1 = \vec{\varphi}_0 + h\mathrm{grad}\psi_0(0, \vec{\varphi}_0),$$

$$\vec{\varphi}_{J-1} = \vec{\varphi}_J + h\mathrm{grad}\psi_1(0, \vec{\varphi}_J),$$

and

$$\triangle_+ f(x_j) = f(x_j + h) - f(x_j), \quad \triangle_- f(x_j) = f(x_j) - f(x_j - h).$$

Suppose that the following conditions are satisfied:

(I) $\psi_0(t, \vec{Z})$ and $\psi_1(t, \vec{Z})$ are two scalar functions which have the continuous derivatives with respect to $t \in [0, T]$, the continuous mixed derivatives of second order with respect to $t \in [0, T]$ and $\vec{Z} \in R^3$. The 3×3 Hessian matrixes $H_0(t, \vec{Z}) = \mathrm{grad\,grad}\psi_0(t, \vec{Z})$ and $H_1(t, \vec{Z}) = \mathrm{grad\,grad}\psi_1(t, \vec{Z})$ are nonnegatively definite, and $\mathrm{grad}\psi_0(t, 0) = \mathrm{grad}\psi_1(t, 0) = 0$.

(II) $\vec{f}_Z(x, t, \vec{Z})$ is semi-bounded, i.e. $\exists b = const$, s.t.

$$\vec{\xi} \cdot \vec{f}_Z(x, t, \vec{Z})\vec{\xi} \le b|\vec{\xi}|^2, \quad \forall \xi \in R^3.$$

(III) $\varphi(x) \in H^2(0, l)$, satisfies the nonlinear boundary condition at the ends of the interval $[0, l]$.

By using the fixed point argument and a priori estimates, we can prove that

Lemma 1.3 Under the conditions (I)-(III), the nonlinear system (1.35)-(1.37) of ordinary differential equation with $\epsilon \ge 0$ has at least one 3-dimensional discrete vector solution $\vec{Z}_h(t) = \{\vec{Z}_j(t)|_{j=0,1,\cdots,J}\}$, $\vec{Z}_j(t) \in C^2([0,T])$, $(j = 1, 2, \cdots, J)$.

Lemma 1.4 Under the conditions (I)-(III), there are relations

$$\sup_{0 \le t \le T} |\triangle_+ \vec{Z}_j(t)| \le K_1 h, \quad \sup_{0 \le t \le T} |\triangle_+^2 \vec{Z}_j(t)| \le K_2 h^{1/2},$$

$$\sup_{0 \le t \le T} |\vec{Z}_j(t + \triangle t) - \vec{Z}_j(t)| \le K_3 \triangle t^{1/2}, \quad t \in [0, T - \triangle t]$$

$$\max_{j=0,\cdots,J} |\triangle_+ \vec{Z}_j(t + \triangle t) - \triangle_+ \vec{Z}_j(t)| \le K_4 h \triangle t^{1/2}, \quad t \in [0, T - \triangle t]$$

where $K_1, \cdots, K_4$ are constants, which are independent of h and dependent on $\epsilon > 0$.

Theorem 1.11 Under the conditions (I)-(III), the nonlinear boundary problem (1.30)-(1.32) $(\epsilon > 0)$ has a unique global solution $\vec{Z}(x, t)$ having the continuous derivatives $\vec{Z}_x(x, t)$, the generalized derivatives $\vec{Z}_{xx}(x, t)$, $\vec{Z}_t(x, t) \in L_\infty(0, T; L_2(0, l))$ and $\vec{Z}_{xt}(x, t)$, $\vec{Z}_{xxx}(x, t) \in L_2(Q_T)$, which satisfy spin system in generalized sense and the nonlinear boundary conditions and initial condition in classical sense.

We are going to construct the global solution $\vec{Z}(x, t)$ of nonlinear boundary problem (1.30)-(1.32) $(\epsilon = 0)$.

Lemma 1.4 Under the conditions (I)-(III), the generalized global vector solution $\overrightarrow{Z}_\epsilon(x,t)$ of nonlinear boundary problem (1.30)-(1.32) ($\epsilon > 0$) has the estimates

$$\|\overrightarrow{Z}_\epsilon\|_{L_\infty(Q_T)} + \sup_{0 \le t \le T} \|\overrightarrow{Z}_{\epsilon x}(\cdot,t)\|_{L_2(0,l)} \le K_5,$$

$$\sup_{0 \le t \le T} \|\overrightarrow{Z}_\epsilon(\cdot,t)\|_{H^{-1}(0,l)} \le K_6,$$

$$|\overrightarrow{Z}_\epsilon(x_1,t) - \overrightarrow{Z}_\epsilon(x_2,t)| \le K_7|x_1 - x_2|^{1/2},$$

$$|\overrightarrow{Z}_\epsilon(x,t_1) - \overrightarrow{Z}_\epsilon(x,t_2)| \le K_8|t_1 - t_2|^{1/4},$$

where K_5, K_6, K_7 and K_8 are constants independent of $\epsilon > 0$ and $x_1, x_2 \in [0,l]$, $t_1, t_2 \in [0,T]$.

Theorem 1.12 Suppose that the conditions (I)-(III) are satisfied, the nonlinear boundary problems (1.30)-(1.32) ($\epsilon = 0$) for the system of ferromagnetic chain has at least one weak solution $\overrightarrow{Z}(x,t) \in L_\infty(0,T;H^1(0,l)) \cap C^{(1/2,1/4)}(Q_T)$.

Now we consider the mixed problems as follows

$$\overrightarrow{Z}_x(0,t) = \mathrm{grad}\psi_0(t, \overrightarrow{Z}(0,t)), \ \ t \ge 0,$$

$$\overrightarrow{Z}(l,t) = 0, \ \ t \ge 0$$

$$\overrightarrow{Z}(x,0) = \varphi(x), \ \ 0 \le x \le l \tag{1.38}$$

for the spin system (1.30) ($\epsilon > 0$) and the system of ferromagnetic chain ($\epsilon = 0$).

Suppose that the conditions (II) and (III) are satisfied. Let us further assume instead of condition (I) the following condition

(I') $\psi_0(t, \overrightarrow{Z})$ is a scalar function, and $\psi_{0t} \in C^0$, $\psi_{0tz} \in C^0$, $\psi_{0ZZ} \in C^0$. The Hessian matrix $H_0(t, \overrightarrow{Z})$ of $\psi_0(t, \overrightarrow{Z})$ with respect to $\overrightarrow{Z} \in R^3$ is nonnegative definite also with $\mathrm{grad}\psi_0(t,0) = 0$, $\varphi(l) = 0$.

Theorem 1.13 Under the conditions (I')(II)(III), the boundary value problem with the mixed nonlinear boundary conditions (1.38) for the spin system (1.30) ($\epsilon > 0$) has a unique generalized solution

$$\overrightarrow{Z}(x,t) \in Z_l = L_\infty(0,T;H^2(0,l)) \cap W_\infty^{(1)}(0,T;L_2(0,l))$$

$$\cap L_2(0,T;H^3(0,l)).$$

By using the estimates that are independent of $\epsilon > 0$ and letting $\epsilon \to 0$, we have

Theorem 1.14 Suppose that the conditions (I')(II)(III) are satisfied, and assume that $\overrightarrow{f}(l,t,0) = 0$. Then the mixed problem (1.38) for the system of ferromagnetic chain (1.30) has at least one weak solution

$$\overrightarrow{Z}(x,t) \in L_\infty(0,T;H^1(0,l)) \cap C^{(1/2,1/4)}(Q_T).$$

Consider the nonlinear boundary problem in the infinite domain $Q_T^* = \{x \in R^+, 0 \leq t \leq T\}$

$$\overrightarrow{Z}_x(0,t) = \mathrm{grad}\psi_0(t, \overrightarrow{Z}(0,t)), \ t \geq 0$$

$$\overrightarrow{Z}(x,0) = \varphi(x), \ 0 \leq x < \infty. \tag{1.39}$$

Suppose that the following additional conditions hold:
(II*) The Jacobi derivative matrix $f_Z \in Q_T^*$, $(Z \in R^3)$ is semi-bounded.
(III*) $\varphi(x) \in H^1(R^+)$.
(IV*)

$$|\overrightarrow{f}(x,t,\overrightarrow{Z})|, |\overrightarrow{f}_x(x,t,\overrightarrow{Z})|, |\overrightarrow{f}_t(x,t,\overrightarrow{Z})| \leq a(x,t)F(\overrightarrow{Z}) + b(x,t),$$

or

$$|\overrightarrow{f}(x,t,\overrightarrow{Z})|, |\overrightarrow{f}_x(x,t,\overrightarrow{Z})|, |\overrightarrow{f}_t(x,t,\overrightarrow{Z})| \leq c(x,t)|\overrightarrow{Z}|^k + d(x,t),$$

where $a(x,t), b(x,t), d(x,t) \in L_\infty(0,T; L_2(R^+))$, $c(x,t) \in L_\infty(Q_T^*)$, $k \geq 0, F(\overrightarrow{Z}) \in C^3$.

Theorem 1.15 Suppose the conditions (I')(II*)(III*) and (IV*) are satisfied. Then the problem (1.39) for the system (1.30) ($\epsilon > 0$) in infinite domain Q_T^* has a unique generalized solution

$$\overrightarrow{Z}(x,t) \in Z_\infty = L_\infty(0,T; H^2(R^+)) \cap W_\infty^{(1)}(0,T; L_2(R^+))$$

$$\cap L_2(0,T; H^3(R^+)).$$

For the system (1.30) ($\epsilon = 0$) in infinite domain Q_T^* has at least one weak solution

$$\overrightarrow{Z}(x,t) \in L_\infty(0,T; H^1(R^+)) \cap C_{loc}^{(1/2,1/4)}(Q_T^*).$$

1.2 Multidimensional Cases

Let Ω be a bounded domain in m-dimensional Enclidean space R^m and have twice continuous differentiable boundary $\partial\Omega$. In the cylindrical domain $Q_T = \{x \in \Omega, 0 \leq t \leq T\}$. Let us consider the homogeneous boundary problem

$$\overrightarrow{Z}(x,t) = 0, \ \text{for } x \in \partial\Omega, \ 0 \leq t \leq T \tag{1.40}$$

wiht the initial value condition

$$\overrightarrow{Z}(x,0) = \varphi(x), \ \text{for } x \in \Omega \tag{1.41}$$

for the system of ferromagnetic chain with several variables

$$\overrightarrow{Z}_t = \overrightarrow{Z} \times \triangle\overrightarrow{Z} + \overrightarrow{f}(x,t,\overrightarrow{Z}), \tag{1.42}$$

where $\overrightarrow{f}(x,t,\overrightarrow{Z})$ is a given 3-dimensional vector function in $x \in R^m, t \in R^+, \overrightarrow{Z} \in R^3$, $\varphi(x)$ is a given 3-dimensional initial value function on $\bar{\Omega}$.

It is difficult to prove the existence of global weak solution by means of Leray-Schauder fixed point argument. We apply the Galerkin method.

Suppose that the following conditions are satisfied:

(I) The 3×3 Jacobi derivative matrix $\partial \overrightarrow{f}(x,t,\overrightarrow{Z})/\partial \overrightarrow{Z}$ of 3 dimensional vector valued function $\overrightarrow{f}(x,t,\overrightarrow{Z})$ with respect to the 3-dimensional vector variable $\overrightarrow{Z}$ is semi-bounded , That is, there exists a constant b such that for any 3-dimensional vector $\xi \in R^3$

$$\overrightarrow{\xi} \cdot \frac{\partial \overrightarrow{f}}{\partial \overrightarrow{Z}} \overrightarrow{\xi} \le b|\overrightarrow{\xi}|^2. \tag{1.43}$$

(II) The system (1.42) is homogeneous, that is $\overrightarrow{f}(x,t,0) = 0$. Furthermore, suppose that $\overrightarrow{f}(x,t,\overrightarrow{Z})$ itself and its derivatives with respect to x have the order growth on $|\overrightarrow{Z}|$ not greater then $1+2/m$, which can be expressed by the following inequality

$$|\overrightarrow{f}(x,t,\overrightarrow{Z})| \le A|\overrightarrow{Z}|^l + B, \ 2 \le l \le 2+4/(m-2) \ (m \ge 2),$$

$$|\nabla \overrightarrow{f}(x,t,\overrightarrow{Z})| \le A|\overrightarrow{Z}|^{1+2/m} + B, \tag{1.44}$$

where A and B are given nonnegative constants and ∇ denotes the partial gradient operator with respect to x. (Here ∇f can be regarded as an $m \times 3$ matrix or tensor).

(III) The components of 3-dimensional initial vector value function $\varphi(x)$ belong to $H_0^1(\Omega)$.

Theorem 1.16 Suppose that the assumptions (I)(II) and (III) are fulfilled. The homogeneous boundary problem (1.40)(1.41) for the homogeneous system (1.42) of ferro-magnetic chain has at least one weak solution

$$\overrightarrow{Z}(x,t) \in L_\infty(0,T;H_0^1(\Omega)) \cap C^{0,\frac{1}{3+[m/2]}}(0,T;L_2(\Omega)).$$

Theorem 1.17 Suppose that the following conditions are fulfilled:

(1) The 3-dimensional vector valued function $\overrightarrow{f}(x,t,\overrightarrow{Z})$ satisfies the inequality

$$\overrightarrow{Z} \cdot \overrightarrow{f}(x,t,\overrightarrow{Z}) \ge C_0|\overrightarrow{Z}|^{2+\delta}$$

for any $(x,t) \in Q_T, Z \in R^3$, where $C_0 > 0, \delta > 0$.

(2) For the initial 3-dimensional vector valued function $\varphi(x)$, we admit $\|\varphi\|_{L_2(\Omega)} > 0$.

Then the 3-dimensional vector valued generalized solution $\overrightarrow{Z} \in W_2^{(2,1)}(Q_T)$ of the system (1.42)(1.40)(1.41) of ferro-magnetic chain blows up in a finite interval of time t, in the sense that $\|\overrightarrow{Z}(\cdot,t)\|_{L_p(\Omega)}$ becomes infinite in the finite value of t for any $2 \le p < \infty$.

1.3   Geometrical extensions for systems of ferromagnetic chain

We consider a class of strongly coupled and strongly degenerate systems of partial differential equations of the form

$$\overrightarrow{Z}_t =^* [g_1(\overrightarrow{Z}) \wedge g_2(\overrightarrow{Z}) \cdots \wedge g_{n-2}(\overrightarrow{Z}) \wedge \triangle \overrightarrow{Z}] + \overrightarrow{f}(x, t, \overrightarrow{Z}), \qquad (1.45)$$

where $\overrightarrow{Z} = (Z_1, Z_2, \cdots, Z_n)(n \geq 2)$ is the n-dimensional unknown vector function of the space variable $x = (x_1, \cdots, x_m) \in R^m$ and the time variable $t \in R^+$. The symbol "$\wedge$" denotes the exterior product operator, "$\triangle$" the Laplacian operator in $R^m (m \geq 1)$ and "$*$" denotes the Hodge star operator. Here $g_k(\overrightarrow{Z})(k = 1, 2, \cdots, n-2)(n \geq 2)$ are the n-dimensional vector functions of variable $\overrightarrow{Z} \in R^m$ and $\overrightarrow{f}(x, t, \overrightarrow{Z})$ is the n-dimensional vector function of variables $x \in R^m, t \in R^+$ and $\overrightarrow{Z} \in R^n$. This system can also be expressed in the form

$$\overrightarrow{Z}_t = A(\overrightarrow{Z})\triangle \overrightarrow{Z} + \overrightarrow{f}(x, t, \overrightarrow{Z}), \qquad (1.46)$$

where $A(\overrightarrow{Z})$ is $n \times n$ matrix of vector variable $\overrightarrow{Z} \in R^n$.

When $n = 2$, (1.45) or (1.46) becomes the component equation of nonlinear Schrodinger equation.

When $n = 3$, (1.45) becomes

$$\overrightarrow{Z}_t = g_1(\overrightarrow{Z}) \times \triangle \overrightarrow{Z} + \overrightarrow{f}(x, t, \overrightarrow{Z}), \qquad (1.47)$$

where "$\times$" denotes the cross product operator of two 3-dimensional vectors. When $g_1(\overrightarrow{Z}) \equiv \overrightarrow{Z}$, this system becomes the well-known system of ferro-manetic chain.

For the general case (1.45) or (1.46), the coefficient matrix $A(\overrightarrow{Z})$ is skew-symmetric and zero definite, the rank of the matrix $A(\overrightarrow{Z})$ equals two. When $n \geq 3$, the determinant $|A(\overrightarrow{Z})|$ equals zero.

This shows that the system (1.45) or (1.46) is strongly coupled and strongly degenerat nonlinear system of partial differential equations, and it is called the system of ferro-magnetic chain type.

Consider three kinds of problems for system (1.45).

(1) We consider first the global solvability of the Cauchy problem $(m = 1)$

$$\overrightarrow{Z}_t =^* [\overrightarrow{Z} \wedge g_2(\overrightarrow{Z}) \wedge \cdots \wedge g_{n-2}(\overrightarrow{Z}) \wedge \overrightarrow{Z}_{xx}] + \overrightarrow{f}(x, t, \overrightarrow{Z}), \qquad (1.48)$$

$$\overrightarrow{Z}(x, 0) = \varphi(x). \qquad (1.49)$$

Suppose that the following conditions are satisfied.

(S_1) $g_k(\overrightarrow{Z})$ $(k = 2, \ldots, n-2)(n \geq 3)$ are continuously differentiable. Vectors $\overrightarrow{Z}, g_2(\overrightarrow{Z}), \ldots, g_{n-2}(\overrightarrow{Z})$ are linearly independent.

(S_2) $\vec{f}(x,t,\vec{Z})$ and $\vec{f}_x(x,t,\vec{Z})$ are continuous for $(x,t,\vec{Z}) \in R \times R^+ \times R^n$, and $\vec{f}_Z(x,t,\vec{Z})$ is semi-bounded, i.e.,

$$\vec{\xi} \cdot \vec{f}_Z(x,t,\vec{Z})\vec{\xi} \le b|\vec{\xi}|^2, \tag{1.50}$$

where $\vec{\xi} \in R^n$ and $(x,t,\vec{Z}) \in R \times R^+ \times R^n$, b is a constant.

(S_3) $\vec{f}(x,t,\vec{Z})$ and $\vec{f}_x(x,t,\vec{Z})$ satisfy

$$|\vec{f}(x,t,\vec{Z})|, |\vec{f}_x(x,t,\vec{Z})| \le a(x,t)F(\vec{Z}) + b(x,t),$$

or

$$|\vec{f}(x,t,\vec{Z})|, |\vec{f}_x(x,t,\vec{Z})| \le c(x,t)|\vec{Z}|^l + d(x,t),$$

where $a(x,t), b(x,t), d(x,t) \in L^\infty(R^+; L_2(R)), c(x,t) \in L^\infty(Q^*_\infty), Q^*_\infty = \{x \in R, t \in R^+\}, l \ge 0$ and $F(\vec{Z})$ is a continuous function of $\vec{Z} \in R^n$.

(S_4) $\varphi(x) = (\varphi_1(x), \ldots, \varphi_n(x)) \in H^1(R)$.

Theorem 1.18 Under the conditions $(S_1)(S_2)(S_3)$ and (S_4), the Cauchy problem $(1.48)(1.49)$ has at least one weak solution $\vec{Z}(x,t) \in L^\infty_{loc}(R^+; H^1(R)) \cap C^{(1/2,1/4)}_{loc}(Q^*_\infty), Q^*_\infty = \{x \in R, t \in R^+\}$, satisfying the initial condition (1.49) in classical sense.

If $b < 0$ and $\vec{f}(x,t,0) \equiv 0$, then

$$\lim_{t\to\infty} \|\vec{Z}(\cdot,t)\|_{L_p(R)} = 0, \ 2 \le p < \infty.$$

If $\vec{f}_x(x,t,\vec{Z}) \equiv 0$, then

$$\lim_{t\to\infty} \|\vec{Z}_x(\cdot,t)\|_{L_2(R)} = 0.$$

(2) Consider the initial boundary value problem for the homogeneous system of the form $(m = 1)$

$$\vec{Z}_t =^* [gradG_1(\vec{Z}) \wedge g_2(\vec{Z}) \wedge \ldots \wedge g_{n-2}(\vec{Z}) \wedge \vec{Z}_{xx}] \tag{1.51}$$

in $Q_T = \{0 \le x \le l, 0 \le t \le T\}$ with any one of the following homogeneous boundary conditions (*)

$$\vec{Z}(0,t) = \vec{Z}(l,t) = 0, \tag{1.52a}$$

$$\vec{Z}_x(0,t) = \vec{Z}_x(l,t) = 0, \tag{1.52b}$$

$$\vec{Z}(0,t) = \vec{Z}_x(l,t) = 0, \tag{1.52c}$$

$$\vec{Z}_x(0,t) = \vec{Z}(l,t) = 0, \tag{1.52d}$$

and the initial value condition

$$\vec{Z}(x,0) = \varphi(x). \tag{1.53}$$

The following assumptions are satisfied.

(S_5) The scalar function $G_1(\vec{Z})$ of $\vec{Z} \in R^n$ is twice continuously differentiable and boundedness of $G_1(\vec{Z})$ implies always the boundedness of $|\vec{Z}|$, i.e.,

$$\lim_{|\vec{Z}| \to \infty} G_1(\vec{Z}) = \infty.$$

(S_6) $g_k(\vec{Z})(k = 2, \ldots, n-2)(n \geq 3)$ are continuously differentiable with respect to $\vec{Z} \in R^n$, $\mathrm{grad} G_1(\vec{Z}), g_2(\vec{Z}), \ldots, g_{n-2}(\vec{Z})$ are linearly independent.

(S_7) $\varphi(x) \in H^1(0,l)$ satisfies one of boundary conditions $(*)$.

By applying the finite difference method it can be proved

Theorem 1.19 Suppose that the conditions $(S_5)(S_6)$ and (S_7) are satisfied. The initial boundary value problem (1.51) (1.52) (1.53) has at least one weak solution $\vec{Z}(x,t) = (Z_1(x,t), \ldots, Z_n(x,t)) \in L^\infty(0,T; H^1(0,l)) \cap C^{(1/2,1/4)}(Q_T)$, satisfying the initial condition (1.53) in classical sense.

(3) In $Q_\infty = \{x \in \bar{\Omega} \subset R^m, 0 \leq t < \infty\}$ we consider the initial boundary value problem

$$\vec{Z}_t = {}^* [\vec{Z} \wedge \vec{a}_2 \wedge \cdots \wedge \vec{a}_{n-2} \wedge \Delta \vec{Z}] + \vec{f}(x,t,\vec{Z}), \tag{1.54}$$

$$\vec{Z}(x,t) = 0, \quad x \in \partial\Omega, 0 \leq t < \infty, \tag{1.55}$$

$$\vec{Z}(x,0) = \vec{\varphi}(x), \quad x \in \bar{\Omega}, \tag{1.56}$$

where $\vec{Z}(x,t) = (Z_1(x,t), \ldots, Z_n(x,t))$ is unknown vector function, $\vec{a}_k (k = 2, \ldots, n-2)$ are linearly independent constant vectors in R^n, $\vec{\varphi}(x) = (\varphi_1(x), \ldots, \varphi_n(x))$ is a known vector function.

The following assumptions are fulfilled.

(S_8) The $n \times n$ Jacobi derivative matrix $\dfrac{\partial \vec{f}(x,t,\vec{Z})}{\partial \vec{Z}}$ is semi-bounded.

(S_9) $\vec{f}(x,t,0) \equiv 0$. Furthermore, suppose that

$$|\vec{f}(x,t,\vec{Z})| \leq A|\vec{Z}|^l + B, |\nabla \vec{f}(x,t,\vec{Z})| \leq A|\vec{Z}|^{1+\frac{2}{m}} + B,$$

where $2 \leq l \leq 2 + \frac{4}{m-2}$ $(m \geq 2)$, A and B are constants.

(S_{10}) The n-dimensional vector function $\varphi(x) \in H_0^1(\Omega)$.

Then, by applying the Galerkin's method we can prove

Theorem 1.20 Suppose that the conditions $(S_8) - (S_{10})$ are satisfied. Then in domain Q_∞ the problem (1.54)-(1.56) has at least one weak solution $\vec{Z}(x,t) \in L_{loc}^\infty(R^+; H_0^1(\Omega)) \cap C_{loc}^{(0, \frac{1}{2+[m/2]})}(R^+; L_2(\Omega))$.

If $\vec{\xi} \cdot \dfrac{\partial \vec{f}(x,t,\vec{Z})}{\partial \vec{Z}} \vec{\xi} \leq b|\vec{\xi}|^2$, and $b < 0$, then we have

$$\lim_{t \to \infty} \|\vec{Z}(\cdot,t)\|_{L_2(\Omega)} = 0.$$

In case $\vec{f} \equiv 0$, if $|\varphi(x)|^2 \equiv const$, and $\vec{a}_{k_i} \cdot \vec{\varphi}(x) \equiv const$, $k_i = 2, \ldots, n-2, i = 0, 1, \ldots, n-3$, then $|\vec{Z}|^2 = const$, and $\vec{a}_{k_i} \cdot \vec{Z} = const.$, this is the (n-s-1)-dimensional hypersphere S^{n-s-1}, $n-s-1 = 2, \ldots, n-1$.

1.4   Landan-Lifshitz equations of the ferro-magnetic spin chain and harmonic maps

Let M be a Riemannian manifold without boundary and N be S^2. We consider the following Landau-Lifshitz type equations of M into S^2

$$\partial_t \vec{Z} = -\alpha_1 \vec{Z} \times (\vec{Z} \times \Delta_M \vec{Z}) + \alpha_2 \vec{Z} \times \Delta \vec{Z}, \qquad (1.57)$$

where $\alpha_1 > 0$ and α_2 are constants.

The heat flow for harmonic map of M into S^2 is given by

$$\partial_t \vec{Z} = \Delta_M \vec{Z} + |\nabla \vec{Z}|^2 \vec{Z}, \qquad (1.58)$$

where the Laplace-Beltrami operator and the norm $|\nabla \vec{Z}|$ in the local coordinate $(x^1, \ldots, x^m)$ on M are denoted by

$$\Delta_M = \frac{1}{\sqrt{\gamma}} \frac{\partial}{\partial x^\beta} (\gamma^{\alpha\beta} \sqrt{\gamma} \frac{\partial}{\partial x^\alpha}) = \gamma^{\alpha\beta} \frac{\partial^2}{\partial x^\alpha \partial x^\beta} - \Gamma^k_{\alpha\beta} \frac{\partial}{\partial x^k} \qquad (1.59)$$

and

$$|\nabla \vec{Z}|^2 = \sum_{\alpha\beta} \sum_i \gamma^{\alpha\beta} \frac{\partial Z_i}{\partial x^\alpha} \frac{\partial Z_i}{\partial x^\beta}. \qquad (1.60)$$

(1) We establish the strong links and relations between harmonic maps and the solution of the ferro-magnetic spin chain.

Theorem 1.21 In the classical sense, $\vec{Z}$ is the solution of equation (1.57) with initial value $\vec{Z}_0(x)$ such that

$$|\vec{Z}_0(x)|^2 = 1, \quad x \in M. \qquad (1.61)$$

If and only if $\vec{Z}$ is the solution of the equation (1.62)

$$\partial_t \vec{Z} = \alpha'_1 \frac{1}{\sqrt{\gamma}} \frac{\partial}{\partial x^\beta} (\gamma^{\alpha\beta} \sqrt{\gamma} \frac{\partial \vec{Z}}{\partial x^\alpha}) + \alpha_1 |\nabla \vec{Z}|^2 \vec{Z}$$

$$+ \alpha_2 \vec{Z} \times \frac{1}{\sqrt{\gamma}} \frac{\partial}{\partial x^\beta} (\gamma^{\alpha\beta} \sqrt{\gamma} \frac{\partial \vec{Z}}{\partial x^\alpha}) \qquad (1.62)$$

with the initial value $\vec{Z}_0(x)$ satisfying (1.61).

Corollary If $\alpha_2 = 0$, the Landau-Lifshitz equations are equivalent to the heat flow (1.58) of harmonic maps.

Theorem 1.22 In the classical sense, $\overrightarrow{Z} : M \to S^2$ is a harmonic map if and only if $\overrightarrow{Z}$ satisfies (1.57) and

$$\frac{\partial \overrightarrow{Z}(x,t)}{\partial t} = 0, \text{ for } t \geq 0.$$

(2) Consider the Landau-Lifshitz type equations from R^2 to S^2. The periodic initial value problem is considered by

$$\partial_t \overrightarrow{Z} = -\alpha_1 \overrightarrow{Z} \times (\overrightarrow{Z} \times \triangle \overrightarrow{Z}) + \alpha_2 \overrightarrow{Z} \times \triangle \overrightarrow{Z} \tag{1.63}$$

with initial value $\overrightarrow{Z}_0(x)$ and the periodic boundary conditions

$$\overrightarrow{Z}(x^1 + L, x^2, t) = \overrightarrow{Z}(x^1 - L, x^2, t),$$

$$\overrightarrow{Z}(x^1, x^2 + L, t) = \overrightarrow{Z}(x^1, x^2 - L, t), \quad x \in R^2, t > 0, \tag{1.64}$$

where L is a positive constant and Q is the domain $\{x = (x^1, x^2) \in R^2, -L \leq x^1 \leq L, -L \leq x^2 \leq L\}$.

Theorem 1.23 Suppose that $\overrightarrow{Z}_0(x)$ is a given initial value in $H^k(Q, S^2)$ satisfying (1.64), where k is large enough. Then there exists a constant λ such that the periodic value problem (1.63)(1.64) with the initial value $\overrightarrow{Z}_0(x)$ has a smooth global solution $\overrightarrow{Z}(x,t)$ provided $\|\nabla \overrightarrow{Z}_0\| \leq \lambda$.

(3) Consider the Landau-Lifshitz equations from a Riemannian surface M into S^2, and establish the existence of a unique global solution to (1.57) for finite initial energy $E(\overrightarrow{Z}_0) < \infty$ which is regular with exception of at most finitely many singular points.

Suppose that M be a closed Riemannian surface, and we denote

$$\vee(M_\tau^T; S^2) = \{\overrightarrow{Z} : M \times [\tau, T] \to S^2 | \overrightarrow{Z} \text{ measurable,}$$

$$\operatorname*{ess\ sup}_{0 \leq t \leq T} \int_M |\nabla \overrightarrow{Z}(\cdot, t)|^2 dM + \int_\tau^T (|\nabla^2 \overrightarrow{Z}|^2 + |\partial_t \overrightarrow{Z}|^2) dM dt < \infty\}.$$

Theorem 1.24 Suppose $\overrightarrow{Z}_1, \overrightarrow{Z}_2 \in \vee(M^T; S^2)$ are solutions to the Landau-Lifshitz equation (1.62) with $\overrightarrow{Z}_1(\cdot, 0) = \overrightarrow{Z}_2(\cdot, 0) = \overrightarrow{Z}_0$. Then $\overrightarrow{Z}_1 = \overrightarrow{Z}_2$ in M^T.

Theorem 1.25 Let $\overrightarrow{Z}_0$ be a smooth map. Then there exists an $\epsilon > 0$ and a map $\overrightarrow{Z} : M \times [0, \epsilon] \to S^2$ with $\overrightarrow{Z} \in L_2^p(M^\epsilon)$ solving the equation

$$\partial_t \overrightarrow{Z} = \alpha_1 \triangle_M \overrightarrow{Z} + \alpha_1 |\nabla \overrightarrow{Z}|^2 \overrightarrow{Z} + \alpha_2 \overrightarrow{Z} \times \triangle_M \overrightarrow{Z} \text{ on } M \times [0, \epsilon],$$

$$\overrightarrow{Z}|_{t=0} = \overrightarrow{Z}_0 \text{ on } M \times \{0\}.$$

Moreover, $\overrightarrow{Z}$ is unique and smooth.

Theorem 1.26 For any initial value $\overrightarrow{Z}_0 \in H^{1,2}(M, S^2)$, there exists a unique solution $\overrightarrow{Z}$ of (1.62) on $M \times [0, \infty)$ which is regular on $M \times [0, \infty)$ with exception of at most finitely many point $(x^k, T^k), 1 \le k \le K$, characterized by the condition that

$$\lim_{T \to T^k} \sup_{T < T^k} E_R(\overrightarrow{Z}(\cdot, t), x^k) > \epsilon_1 \text{ for all } R \in (0, R_0],$$

where $E_R(\overrightarrow{Z}, x) = \int_{B_R^M(x)} e(\overrightarrow{Z}) dM, B_R^M(x) = \{y \in M, |x - y|_M < R\}, e(\overrightarrow{Z}) = \frac{1}{2} |\nabla \overrightarrow{Z}|^2$, and ϵ_1 is a constant depending on Riemannian manifold M.

(4) Suppose that M be a compact m-dimensional Riemannian manifold without boundary and $m \ge 3$.

Definition A vector function $\overrightarrow{Z}(x, t)$ is called a global weak solution to (1.57) (1.61), if $\overrightarrow{Z}$ is defined a.e. in $R^+ \times M$ such that

(i) $\overrightarrow{Z}(x, t) \in L^\infty(0, \infty; H^{1,2}(M)), \partial_t \overrightarrow{Z} \in L_2(0, \infty; L_2(M))$,

$$\nabla \cdot (\overrightarrow{Z} \times \nabla \overrightarrow{Z}) \in L_{loc}^2(0, \infty; L_2(M));$$

(ii) $|\overrightarrow{Z}(x, t)|^2 = 1$, a.e. on $R^+ \times M$;

(iii) For any $T < \infty$, $\overrightarrow{Z}$ satisfies the following equality

$$\int_0^T \int_M [\partial_t \overrightarrow{Z} \cdot \varphi + \alpha_1(\overrightarrow{Z} \times \nabla \cdot (\overrightarrow{Z} \times \nabla)\varphi - \alpha_2 \nabla \cdot (\overrightarrow{Z} \times \nabla \overrightarrow{Z})\varphi] dM dt = 0,$$

for all $\varphi \in L_2(0, T; W^{1,2}(M))$;

(iv) $\overrightarrow{Z}(x, 0) = \overrightarrow{Z}_0(x)$ in the trace sense.

Theorem 1.27 Let M be a compact smooth m-dimendional Riemannian manifold without boundary. Let $\overrightarrow{Z}_0 : M \to S^2$ satisfing $\overrightarrow{Z}_0(x) \in H^{1,2}(M)$. Then there exists a global weak solution $\overrightarrow{Z}(x, t)$ to (1.57) (1.61) such that $\overrightarrow{Z}(x, t)$ satisfies

$$\int_0^T \int_M |\partial_t \overrightarrow{Z}|^2 dM dt + \int_M |\nabla \overrightarrow{Z}|^2 dM + \int_0^T \int_M |\nabla \cdot (\overrightarrow{Z} \times \nabla \overrightarrow{Z})|^2 dM dt \le C$$

for each $T \le \infty$, where C is a constant depending on T.

2 Initial value problems for a nonlinear singular integral- differential equation of deep water

The equation which discribes the propagation of intenal waves in the stratified fluid of finite depth is first derived by R. I. Joseph and can be expressed in the form

$$u_t + 2uu_x + Gu_{xx} = 0, \tag{2.1}$$

where $G(\cdot)$ is a singular integral operator defined by

$$Gu(x, t) = \frac{\lambda}{2} P \int_{-\infty}^{\infty} [cosh \frac{\lambda \pi}{2}(y - x) - sgn(y - x)] u(y, t) dy,$$

$\frac{1}{\lambda}$ is the parameter characterizing the depth of fluid and P denotes the principle value of integral. For the shallow water limit as $\lambda \to \infty$, this equation redeuces to the well-known Korteweg-de Vries equation.

$$u_t + 2uu_x + u_{xxx} = 0. \tag{2.2}$$

For the deep water limit, the equation (2.1) reduces to the following form

$$u_t + 2uu_x + Hu_{xx} = 0, \tag{2.3}$$

where H is the Hilbert transform

$$Hu(x,t) = \frac{1}{\pi} P \int_{-\infty}^{\infty} \frac{u(y,t)}{y-x} dy. \tag{2.4}$$

The equation (2.1) and (2.3) are the nonlinear partial differential equation with the singular integral operator. The equation (2.3) is the equation of deep water and is usually called the Benjamin-ono equation.

The study of these equations has the great interest in the physical and mathematical point of view. For example, there is a great deal of work contributed to the soliton solutions and the behaviors of the solutions of the problems for the Korteweg-de Vries equations and their various generalizations.

Very recently there are many investigations of the physical purpose for the nonlinear partial differential equation (2.3) of deep water. The Backlund transformations, the conservation laws, various soliton solutions and their interactions for the Benjamin-Ono equation (2.3) are studied in [15-17]. A lot of work ([18-20]) contributed to the study of the existence, uniqueness and regularity on equation of Benjamin-Ono type.

If the effect of the amplitude of the internal wave is taken into account in the deep fluid, the equation (2.3) has an additional linear term as follows

$$u_t + c_0 u_x + 2uu_x + Hu_{xx} = 0. \tag{2.5}$$

We consider the following initial value problems of the nonlinear singular integral-differential equation of deep water

$$u_t + 2uu_x + Hu_{xx} + b(x,t)u_x + c(x,t)u = f(x,t), \tag{2.6}$$

$$u|_{t=0} = \varphi(x). \tag{2.7}$$

The existence and uniqueness theorems of the generalized and classical global solution for the initial value problems of the Benjamin-Ono equation are proved. The solutions of the mentioned problems are approximated by the solutions of the initial value problems for the equation

$$u_t + 2uu_x + Hu_{xx} - \epsilon u_{xx} + b(x,t)u_x + c(x,t)u = f(x,t), \tag{2.8}$$

obtained by increasing a diffusion term ϵu_{xx} with small coefficient to the original equation (2.6). Singular integral-differential equation (2.6) is builded up by the limiting process of the vanishing of the diffusion coefficient $\epsilon \to 0$. The estimations of the convergence speed are made in the order of the diffusion coefficient ϵ are also obtained.

Lemma 2.1 Suppose that $b(x,t), b_x(x,t), c(x,t) \in L_\infty(Q_T^*), Q_T^* = \{x \in R, 0 \le t \le T\}$, and $f(x,t) \in L_2(Q_T^*)$ and assume that $\varphi(x) \in L_2(R)$. The generalized global solution $u_\epsilon(x,t) \in W_2^{(2,1)}(Q_T^*)$ of the initial value problem (2.7) for the nonlinear parabolic equation (2.8) with Hilbert operator has the estimation

$$\sup_{0 \le t \le T} \|u_\epsilon(\cdot,t)\|_{L_2(R)} \le K_1\{\|\varphi\|_{L_2(R)} + \|f\|_{L_2(Q_T^*)}\}, \tag{2.9}$$

where K_1 is a constant independent of $\epsilon > 0$, but dependent on the norms $\|b_x\|_{L_\infty(Q_T^*)}$ and $\|c\|_{L_\infty(Q_T^*)}$.

There is the equality

$$\frac{d}{dt} \int_{-\infty}^{\infty} (u^4 + 2bu^3 + 3buHu_x + 3u^2Hu_x + 2u_x^2)dx$$

$$= -4\epsilon\|u_{xx}(\cdot,t)\|_{L_2(R)}^2 + \epsilon \int_{-\infty}^{\infty} \{6(uHu_x + H(uu_x)) + 3(bHu_x + H(bu_x)_x)$$

$$+4u^3 + 6bu^3\}u_{xx}dx + \int_{-\infty}^{\infty} [(\frac{5}{2}b_x - 4c)u_x^2 + \frac{3}{2}b_x(Hu_x)^2 - 3b^2u_xHu_x]dx$$

$$-16\int_{-\infty}^{\infty} b^3 u_x dx + 6\int_{-\infty}^{\infty} [(b_x - c)(u^2Hu_x + uH(uu_x)) - b^2u^2u_x]dx$$

$$+3\int_{-\infty}^{\infty} [-bc + b_t)uHu_x + (b_{xx} - 4c_x)uu_x - bu_xH(b_xu)_cuH(bu_x)]dx$$

$$+6\int_{-\infty}^{\infty} f(uHu_x + H(uu_x))dx + \int_{-\infty}^{\infty} [3bHu_x + 3fH(bu_x) + 4u_xf_x]dx$$

$$-\int_{-\infty}^{\infty} cu^4 dx + \int_{-\infty}^{\infty} (2b_t - 6bc)u^3 dx + 4\int_{-\infty}^{\infty} fu^3 dx + 6\int_{-\infty}^{\infty} fu^2 dx$$

$$-3\int_{-\infty}^{\infty} cuH(b_xu)dx + 3\int_{-\infty}^{\infty} fH(b_xu)dx. \tag{2.10}$$

Lemma 2.2 Suppose that $b(x,t) \in W_\infty^{(2,1)}(Q_T^*)$, $c(x,t) \in W_\infty^{(1,0)}(Q_T^*)$ and $f(x,t) \in W_2^{(1,0)}(Q_T^*)$ and assume that $\varphi(x) \in H^1(R)$. The generalized global solutions $u_\epsilon(x,t) \in W_2^{(2,1)}(Q_T^*)$ of the initial value problem (2.7) for the nonlinear parabolic equation (2.8) with Hilbert operator has the estimation

$$\sup_{0 \le t \le T} \|u_{\epsilon x}(\cdot,t)\|_{L_2(R)} \le K_2\{\|\varphi\|_{H^1(R)} + \|f\|_{W_2^{(1,0)}(Q_T^*)}\}, \tag{2.11}$$

where K_2 is a constant independent of $\epsilon > 0$.

There is the equality

$$\frac{d}{dt} \int_{-\infty}^{\infty} (2u_{xx}^2 + 5u_x^2Hu_x + 10uu_xHu_{xx})dx + 4\epsilon\|u_{xxx}(\cdot,t)\|_{L_2(R)}^2$$

$$= -10\epsilon \int_{-\infty}^{\infty} [uHu_{xxx} + H(uu_x)_{xx} + (u_xHu_x)_x - H(u_xu_{xx})]u_{xx}dx$$

$$- \int_{-\infty}^{\infty} [(6b_x + 4c)u_{xx}^2 + 10(bu + 3u^2)u_{xx}Hu_{xx} + 10bu_{xx}H(uu_{xx})]dx$$

$$+ 10 \int_{-\infty}^{\infty} [(2u + b)u_xHu_x - bHu_x^2 + 2u_xH(uu_x)]u_{xx}dx$$

$$- 20 \int_{-\infty}^{\infty} uu_x^2Hu_{xx}dx - 10 \int_{-\infty}^{\infty} bu_xH(u_xu_{xx})dx$$

$$+ \int_{-\infty}^{\infty} [10cuHu_x - 4(2b_{xx} + c_x)u_x]u_{xx}dx - 10 \int_{-\infty}^{\infty} (b_x - c)uu_xHu_{xx}dx$$

$$- 10 \int_{-\infty}^{\infty} (b_x + c)u_xH(uu_{xx})dx - 10 \int_{-\infty}^{\infty} cuH(u_xu_{xx})dx$$

$$+ 10 \int_{-\infty}^{\infty} [H(u_xu_{xx}) - u_{xx}Hu_x]fdx - 4 \int_{-\infty}^{\infty} (c_{xx}u - f_{xx})u_{xx}dx$$

$$- 10 \int_{-\infty}^{\infty} (c_xu^2 - f_xu)Hu_{xx}dx - 10 \int_{-\infty}^{\infty} (c_xu + f_x)H(uu_{xx})dx$$

$$- 10 \int_{-\infty}^{\infty} (b_x + c)u_xHu_x^2dx - 10 \int_{-\infty}^{\infty} (c_xu - f_x)H(u_x^2)dx. \tag{2.12}$$

Lemma 2.3 Suppose that $b(x,t) \in W_\infty^{(2,1)}(Q_T^*)$, $c(x,t) \in W_\infty^{(2,0)}(Q_T^*)$ and $f(x,t) \in W_2^{(2,0)}(Q_T^*)$, and assume that $\varphi(x) \in H^2(R)$. Then the generalized global solution $u_\epsilon(x,t) \in W_2^{(2,1)}(Q_T^*)$ of the initial value problem (2.7) for the nonlinear parabolic equation (2.8) with Hilbert operator has the estimation

$$\sup_{0 \le t \le T} \|u_{\epsilon xx}(\cdot, t)\|_{L_2(R)} \le K_3\{\|\varphi\|_{H^2(R)} + \|f\|_{W_2^{(2,0)}(Q_T^*)}\}, \tag{2.13}$$

where the constant K_3 depends on the norms $\|b\|_{W_\infty^{(2,1)}(Q_T^*)}$, $\|c\|_{W_\infty^{(2,0)}(Q_T^*)}$, $\|\varphi\|_{H^1(R)}$ and independent of $\epsilon > 0$.

Lemma 2.4 Under the conditions of Lemma 2.3 , the set $\{u_\epsilon(x,t)\}$ of the generalized global solutions of initial value problem (2.8) and (2.7) has the following estimations

$$|u_\epsilon(\bar{x},t) - u_\epsilon(x,t)| \le K_4|\bar{x} - x|, \tag{2.14}$$

$$|u_\epsilon(x,\bar{t}) - u_\epsilon(x,t)| \le K_5|\bar{t} - t|^{\frac{3}{4}}, \tag{2.15}$$

$$|u_{\epsilon x}(\bar{x},t) - u_{\epsilon x}(x,t)| \le K_6|\bar{x} - x|^{\frac{1}{2}}, \tag{2.16}$$

and

$$|u_{\epsilon x}(x,\bar{t}) - u_{\epsilon x}(x,t)| \le K_7|\bar{t} - t|^{\frac{1}{4}}, \tag{2.17}$$

where $\bar{x}, x \in R, \bar{t}, t \in [0,T]$ and the constants $K_4 - K_7$ are independent of $\epsilon > 0$.

Passing to the limit as $\epsilon \to 0$, we get

Theorem 2.1 Suppose that $b(x,t) \in W_\infty^{(2,1)}(Q_T^*)$, $c(x,t) \in W_\infty^{(2,0)}(Q_T^*)$ and $f(x,t) \in W_2^{(2,0)}(Q_T^*)$ and also suppose that $\varphi(x) \in H^2(R)$. For the initial value problem (2.6)(2.7) there exists at least one generalized global solution $u(x,t) \in Z = L_\infty(0,T;H^2(R)) \cap W_\infty^1(0,T;L_2(R))$, which satisfies the equation (2.6) in generalized sense and satisfies the initial condition (2.7) in classical sense.

Theorem 2.2 Suppose that $b(x,t) \in W_\infty^{(1,0)}(Q_T^*)$ and $c(x,t) \in L_\infty(Q_T^*)$. The generalized global solution $u(x,t) \in Z$ for the initial value problem (2.6)(2.7) is unique.

Theorem 2.3 Under the conditions of Theorem 2.1, for the generalized global solutions $u_\epsilon(x,t) \in W_2^{(2,1)}(Q_T^*)$ and $u(x,t) \in Z$ of the initial problem (2.7) for the nonlinear parabolic equation (2.8) and the nonlinear integral-differential equation (2.6) respectively, there are the estimations for the rate of convergence in term of the power of the diffusion coefficient $\epsilon > 0$ as follows

$$\sup_{0 \le t \le T} \|u_\epsilon(\cdot,t) - u(\cdot,t)\|_{L_2(R)} \le K_8\epsilon, \tag{2.18}$$

$$\|u_\epsilon - u\|_{L_2(Q_T^*)} \le K_9\epsilon^{\frac{1}{4}}, \tag{2.19}$$

$$\sup_{0 \le t \le T} \|u_{\epsilon x}(\cdot,t) - u_x(\cdot,t)\|_{L_2(R)} \le K_{10}\epsilon^{\frac{1}{2}}, \tag{2.20}$$

and

$$\|u_{\epsilon x} - u_x\|_{L_2(Q_T^*)} \le K_{11}\epsilon^{\frac{1}{4}}, \tag{2.21}$$

where $K_8 - K_{11}$ are the constants independent of $\epsilon > 0$.

Theorem 2.4 Suppose that $\varphi(x) \in H^M(R)$ for $M \ge 2$. The initial value problem (2.7) for the Benjamin-Ono equation (2.3) has a unique global solution $u(x,t) \in \cap_{k=0}^{[\frac{M}{2}]} W_{\infty,loc}^k(R^+;H^{M-2k}(R))$, which has the derivatives $u_{x^r t^s}(x,t) \in L_{\infty,loc}(R^+;L_2(R))$ for $0 \le 2s + r \le M$.

We are going to consider the Cauchy problem for the equation of finite-depth fluids([21])

$$u_t + 2uu_x - G(u_{xx}) = 0, \tag{2.22}$$

where

$$G(u) = P \int_{-\infty}^{\infty} \frac{1}{2\delta}(\cosh\frac{\pi(x-y)}{2\delta} - sgn(x-y))u(y)dy \tag{2.23}$$

is a singular integral operator. We also consider the generalized equation of finite-depth fluids with diffusion term

$$U_t = \alpha U_{xx} + \beta G(U_{xx}) + \varphi_x(U) \tag{2.24}$$

where $\varphi(\cdot)$ is assumed to be a mildly smooth function on R such that

$$|\varphi^{(j)}(u)| \le C(1 + |u|^{3-j}) \text{ for } j = 0,1, u \in R, \tag{A}$$

where α, β, δ are constants with $\alpha \ge 0, \delta > 0$. Equation (2.22) appears in the studying of oceanics and atmospheric science which describes the evolution of long internal waves with small amplitude in a stably stratified, propagating in one direction.

Theorem 2.5 Let $u_0(x)$ be given in $H^{k+1}(R)$, $\varphi(u) \in C^{k+2}$ satisfy the condition (A), the constant $\alpha > 0$. Then the Cauchy problem of the generalized equation (2.24) with diffusion term and with the initial condition

$$u|_{t=0} = u_0(x) \tag{2.25}$$

has a unique global solution $u(x,t) \in W_2^{k+2,\lceil\frac{k+2}{2}\rceil}(Q_T) \cap W_{\infty,2}^{k+2,\lceil\frac{k+2}{2}\rceil}(Q_T)$, for $k \geq 0$.

Theorem 2.6 Let the initial data $u_0(x)$ be given in $H^1(R)$. Then the Cauchy problem of the equation of finite-depth fluids (2.22)(2.25) has at least one global weak solution

$$u(x,t) \in L^\infty(0,\infty; H^1(R)) \cap C^{(\frac{1}{2},\frac{1}{4})}(Q_\infty).$$

Now we consider the long time behavior for the finite-depth fluids

$$u_t - G(u_{xx}) - (\frac{u^p}{p})_x = 0, \tag{2.26}$$

where p is an integer larger than 1.

Theorem 2.7 Let $\delta \in (0,\infty)$, $q = 2p$ and $p > \frac{5}{2} + \sqrt{\frac{21}{2}}$. Assume that the initial data $u_0(x)$ is sufficiently small in $H^3(R) \cap W^{2,\frac{2p}{3p-1}}(R)$. Then the solution of the Cauchy problem (2.26)(2.25) such that

$$\|u(t)\|_{W^{2,q}(R)} \leq C(1+t)^{-\frac{1}{3}(1-\frac{2}{q})} \tag{2.27}$$

for all $t \geq 0$, where the constant C is independent of u and t.

Theorem 2.8 Under the conditions of Theorem 2.7, the solution of the nonlinear problem (2.26)(2.25) is freely asympotic to the solution of linear problem

$$u_t - G(u_{xx}) = 0, \quad u(x,0) = u_0(x). \tag{2.28}$$

We are concerned with the existence, uniqueness and convergence of solution for the following Cauchy problem [22]

$$u_t + \beta u_{xxx} + \delta H u_{xx} + \gamma H u_x + g(u)_x = 0, \tag{2.29}$$

$$u(x,0) = \varphi(x), \tag{2.30}$$

for $x \in R$, $t \in R^+$, $k \in N$, where β, δ, γ are real constants, $H(u) = \frac{1}{\pi} P. \int_{-\infty}^\infty \frac{u(y,t)}{y-x} dy$ is the Hilbert transform. It is known that Eq. (2.29) plays an important rule in many branches of physics. For instance, if $g(u) = cu + u^2$ and $\gamma = 0$, then eq. (2.29) reduces to Benjamin- Ono-KdV equation

$$u_t + \beta u_{xxx} + \delta H u_{xx} + g(u)_x = 0, \tag{2.31}$$

which describes a large class of internal waves in the ocean and stratified fluid. Moreover, if $\delta = 0$ or $\beta = 0, c = 0$, then the BO-KdV equation (2.31) reduces respectively to the well known Korteweg-de Vries equation and Benjamin-Ono equation. When $\delta = 0$, eq. (2.31) reduces to the following equation

$$u_t + \beta u_{xxx} + \gamma H u_x + g(u)_x = 0, \tag{2.32}$$

which is the general equation obtained by Ott and Sudam [23] in studying ionacoustic waves of finite amplitude with the linear Landau damping.

Theorem 2.9 Let β, δ, γ be constants satisfying $\beta \neq 0, \gamma \leq 0$. Let initial function $\varphi(x) \in H^k(R)$, and $g(u) \in C^k(R)$ $(k \geq 0)$. If there exists a constant C such that

$$|g'(v)| \leq C(1 + |v|^\eta), \qquad (2.33)$$

where $0 < \eta \leq 2$, then the Cauchy problem (2.30) in domain Q_T for the nonlinear dispersive equation (2.29) has a unique global classical solution $u \in W_\infty^r(0, T; H^{k-3r}(R))$ with $r \leq [k/3]$.

Theorem 2.10 Let $u_{\delta\gamma}(x, t) \in W_\infty^r(0, T; H^{k-3r}(R))$ be the solution of problem (2.29)(2.30) obtained in Theorem 2.8. Then there exist three functions u_δ, u_γ and u such that

(a) $\lim_{\delta \to 0} u_{\delta\gamma}(x, t) = u_\gamma(x, t)$ and $u_\gamma(x, t) \in \cap_{r=0}^{[k/3]} W_\infty^r(0, T; H^{k-3r}(R))$ is the unique global smooth solution of problem (2.32) (2.30).

(b) $\lim_{\gamma \to 0} u_{\delta\gamma}(x, t) = u_\delta(x, t)$ and $u_\delta(x, t) \in \cap_{r=0}^{[k/3]} W_\infty^r(0, T; H^{k-3r}(R))$ is the unique global smooth solution of problem (2.31) (2.30).

(c) $\lim_{\delta, \gamma \to 0} u_{\delta\gamma}(x, t) = \lim_{\delta \to 0} u_\delta(x, t) = \lim_{\gamma \to 0} u_\gamma(x, t) = u(x, t)$ and $u(x, t) \in \cap_{r=0}^{[k/2]} W_\infty^r(0, T; H^{k-2r}(R))$ is the unique global smooth solution of problem

$$u_t + \beta u_{xxx} + g(u)_x = 0,$$

$$u|_{t=0} = \varphi(x).$$

3   On the System of Korteweg-de Vries Equations

In recent years, many authors [24-27] have studied the properties of the solutions for the Korteweg-de Vires equation

$$u_t + \alpha u u_x + \beta u_{xxx} = 0, \qquad (3.1)$$

and its generalized type, where α and β are constants. The KdV equation

$$Z_t + \alpha(|Z|^2 Z)_x + Z_{xxx} = 0, \qquad (3.2)$$

has been studied in [28], where $Z(x, t) = u(x, t) + iv(x, t)$ is a complex function. The equation (3.2) can be written in the form of system of the real functions $u(x, t)$ and $v(x, t)$

$$u_t + \alpha((u^2 + v^2)u)_x + u_{xxx} = 0,$$
$$v_t + \alpha((u^2 + v^2)v)_x + v_{xxx} = 0. \qquad (3.3)$$

We consider the following system of generalized KdV type [29-31]

$$\vec{u}_t + (\mathrm{grad}\varphi(\vec{u}))_x + \vec{u}_{xxx} = \vec{f}(x, t, \vec{u}), \qquad (3.4)$$

where $\vec{u}(x, t) = (u_1(x, t), \ u_2(x, t), \ \cdots, u_J(x, t))$ is a J dimensional functional vector, $\varphi(\vec{u}) = \varphi(u_1, \cdots, u_J)$ is a scalar function of vector $\vec{u}$, $\vec{f}(x,$

$t, \overrightarrow{u}) = (f_1(x, t, u_1, \cdots, u_J), \cdots, f_J(x, t, u_1, \cdots, u_J))$ is a J dimensional functional vector. For the system (3.4), we discuss the periodic boundary problem in domain $Q_T = \{(x, t) | -\infty < x < \infty, 0 \le t \le T\}$,

$$\overrightarrow{u}(x, 0) = \overrightarrow{u}_0(x), \quad -\infty < x < \infty$$

$$\overrightarrow{u}(x + 2D, t) = \overrightarrow{u}(x, t), \quad -\infty < x < \infty, \ 0 \le t \le T \tag{3.5}$$

where $\overrightarrow{u}_0(x) = (u_{01}, \cdots, u_{0J}(x))$ is a J dimensional periodic functional vector with period $2D$ and $D > 0$ is a constant. We also consider the initial value problem

$$\overrightarrow{u}(x, 0) = \overrightarrow{u}_0(x), \quad -\infty < x < \infty \tag{3.6}$$

where $\overrightarrow{u}_0(x)$ is a J dimensional functional vector $x \in (-\infty, \infty)$.

First, we study the global generalized solutions and the global classical solutions of periodic boundary problem for the system with small parameter

$$u_t + \epsilon \overrightarrow{u}_{xxxx} + (\mathrm{grad}\varphi(\overrightarrow{u}))_x + \overrightarrow{u}_{xxx} = \overrightarrow{f}(x, t, \overrightarrow{u}), \tag{3.7}$$

where $\epsilon > 0$. Then we prove that the solution $\overrightarrow{u}_\epsilon(x, t)$ of problem (3.7) (3.5) tends to the global solution $\overrightarrow{u}(x, t)$ of the periodic boundary problem for the corresponding system (3.4) as $\epsilon \to 0$.

By using Leray-Schauder fixed point argument, one can prove

Theorem 3.1 Suppose that the following conditions are satisfied for the problem (3.7)(3.5)

(1) $\|\overrightarrow{u}_0\|_{W_2^{(2)}(-D, D)} < \infty$,

(2) $\overrightarrow{f}(x, t, \overrightarrow{u}) \in C^1$, $\overrightarrow{f}(x, t, 0) \in L_2(Q_T)$, $\overrightarrow{f}(x, t, \overrightarrow{u})$ is a periodic functional vector for x with period $2D$, and there is the estimation

$$|\overrightarrow{f}_x(x, t, \overrightarrow{u})| \le c(x, t)|u|^3 + d(x, t),$$

where $c(x, t)$ is bounded in Q_T, $d(x, t) \in L_2(Q_T)$.

(3) For the functional vector $\overrightarrow{f}(x, t, \overrightarrow{u})$, the Jacobi derivatives matrix $\partial \overrightarrow{f}/\partial u = \overrightarrow{f}_u(x, t, \overrightarrow{u})$ is semi-bounded, i.e., there exists a constant b such that for any J dimensional vector $\xi \in R^J$,

$$(\xi, \frac{\partial \overrightarrow{f}}{\partial \overrightarrow{u}}\xi) \le b(\xi, \xi).$$

(4) $\phi(\overrightarrow{u}) \in C^3$, and

$$|\phi(\overrightarrow{u})| \le C|\overrightarrow{u}|^l, \ |\mathrm{grad}\varphi(\overrightarrow{u})| \le C|\overrightarrow{u}|^{l-1},$$

$$|\frac{\partial^2 \phi(\overrightarrow{u})}{\partial u_i \partial u_j}| \le C(|\overrightarrow{u}|^{l-2} + 1), \ (i, j = 1, 2, \cdots, J),$$

$$|(\mathrm{grad}\phi(\overrightarrow{u}), \overrightarrow{f}(x, t, \overrightarrow{u}))| \le C|\overrightarrow{u}|^l, \tag{Φ}$$

where $l = 6 - \delta, \delta > 0$ and C is a constant. Then there exists a unique global generalized solution $\vec{u}(x,t)$ for the problem (3.7)(3.5), and

$$\vec{u}(x,t) \in L_\infty(0,T;W_2^{(2)}(-D,D)) \cap W_2^{(1)}(0,T;L_2(-D,D))$$

$$\cap L_2(0,T;W_2^{(1)}(-D,D)).$$

Corollary 3.1 Suppose that the conditions of Theorem 3.1 are satisfied and assume that for $k \geq 0, h \geq 0$

(1)' $\vec{u}_0(x) \in W^{(2(2h+1)+h)}(-D,D)$,

(2)' $\vec{f}(x,t,\vec{u}) \in C^{(k+1,h,k+1)}$,

(3)' $\phi(\vec{u}) \in C^{(k+h+3)}$.

Then there exists a smooth global solution $\vec{u}(x,t)$ of the problem (3.7)(3.5) and

$$D_x^h D_t^k \vec{u}(x,t) \in L_\infty(0,T;W_2^{(1)}(-D,D)) \cap W_2^{(1)}(0,T;L_2(-D,D))$$

$$\cap L_2(0,T;W_2^{(4)}(-D,D)).$$

Lemma 3.1 Assume that the conditions of Theorem 3.1 are satisfied. Then for the solution of problem (2.7)(2.5), we have the estimation

$$\sup_{0 \leq t \leq T} \|\vec{u}_\epsilon(\cdot,t)\|_{W_2^{(1)}(-D,D)} \leq K,$$

where K is a constant independent of ϵ and D. So

$$\sup_{-D \leq x \leq D} |\vec{u}_\epsilon(\cdot,t)| \leq K.$$

From the equlity

$$3(\vec{u}_{xx}, \vec{u}_{xx})_t - 5((\mathrm{grad}\varphi(\vec{u}))_x, \vec{u}_x) + 6\epsilon(\vec{u}_{xxxx}, \vec{u}_{xxxx})$$

$$+[6\epsilon(\vec{u}_{xx}, \vec{u}_{xxxx}) - 6\epsilon(\vec{u}_{xxx}, \vec{u}_{xxxx}) + 6(\vec{u}_{xx}, \vec{u}_{xxxx})$$

$$-3(\vec{u}_{xxx}, \vec{u}_{xxx}) + 8\sum_{i,j} \varphi_{ij} \vec{u}_{ixx} \vec{u}_{jxx}$$

$$-10\sum_{i,j} \varphi_{ij} \vec{u}_{ix} \vec{u}_{jxxx} + 5\sum_{i,j,k} \varphi_{ijk} \vec{u}_{ix} \vec{u}_{jx} \vec{u}_{kxx}]_x$$

$$= 6(\vec{u}_{xx}, D_x^2 \vec{f}) - 10\sum_{i,j} \varphi_{ij} \vec{u}_{ix} D_x \vec{f}_j - 5\sum_{i,j,k} \varphi_{ijk} \vec{u}_{ix} \vec{u}_{jx} \vec{f}_k$$

$$+5\epsilon\sum_{i,j,k} \varphi_{ijk} \vec{u}_{ix} \vec{u}_{jx} \vec{u}_{jxxxx} + 10\epsilon\sum_{ij} \varphi_{ij} \vec{u}_{ix} \vec{u}_{jxxxxx}$$

$$-\sum_{i,j,k,l} \varphi_{ijkl} \vec{u}_{ixx} \vec{u}_{jx} \vec{u}_{kx} \vec{u}_{lx}$$

$$+15\sum_{i,j,k,l} \varphi_{ijk}\varphi_{kl} \vec{u}_{ix} \vec{u}_{jx} \vec{u}_{lx} + 10\sum_{i,j,k} \varphi_{ij}\varphi_{jk} \vec{u}_{ix} \vec{u}_{kxx},$$

we can get

Lemma 3.2 Assume that the conditions of Theorem 3.1 are satisfied and also

(2)" $\overrightarrow{f}(x, t, \overrightarrow{u}) \in C^{(2,0,2)}$,

(3)" $\varphi(\overrightarrow{u}) \in C^4$.

Then for the solution $\overrightarrow{u}_\epsilon(x, t)$ of problem (3.7)(3.5), we have

$$\sup_{0 \leq t \leq T} \|\overrightarrow{u}_{\epsilon x x}(\cdot, t)\|_{L_2(-D, D)} \leq K,$$

where the constant K is independent of ϵ and D.

Theorem 3.2 Suppose that the following conditions for the problem (3.4)(3.5) are satisfied:

(1) $\overrightarrow{u}_0(x)$ is a J dimensional functional vector with period $2D$, and $\overrightarrow{u}_0(x) \in W_2^{(5)}(-D, D)$,

(2) $\overrightarrow{f}(x, t, \overrightarrow{u}) \in C^{(4,1,4)}$, $\overrightarrow{f}(x, t, 0) \in L_2(Q_T)$, $\overrightarrow{f}(x, t, \overrightarrow{u})$ is a periodic functional vector for x with period $2D$, and there is the estimate

$$|\overrightarrow{f}_x(x, t, \overrightarrow{u})| \leq c(x, t)|\overrightarrow{u}|^3 + d(x, t),$$

where $c(x, t)$ is bounded in Q_T, $d(x, t) \in L_2(Q_T)$,

(3) For the functional vector $\overrightarrow{f}(x, t, \overrightarrow{u})$, the Jacobi derivative matrix $\overrightarrow{f}_u(x, t, \overrightarrow{u})$ is semi-bounded.

(4) $\phi(\overrightarrow{u}) \in C^5$, and conditions (Φ).

Then there exists a unique global generalized solution of problem (3.4)(3.5), and

$$\overrightarrow{u}(x, t) \in L_\infty(0, T; W^{(4)}(-D, D)) \cap W_\infty^{(1)}(0, T; W_2^{(2)}(-D, D)).$$

As a result, u, u_x, u_{xx} and u_{xxx} are Holder continuous in Q_T.

Theorem 3.3 Suppose that the conditions of Theorem 3.2 are satisfied. Then the global generalized solution $\overrightarrow{u}(x, t)$ of problem (3.4)(3.5) in $L_\infty(0, T; W_\infty^{(2)}(-D, D))$ as $\epsilon \to 0$, and

$$\|\overrightarrow{u}_\epsilon(\cdot, t) - \overrightarrow{u}(\cdot, t)\|_{L_\infty(Q_T)} = O(\epsilon^{7/8}),$$

$$\|\overrightarrow{u}_{\epsilon x}(\cdot, t) - \overrightarrow{u}_x(\cdot, t)\|_{L_\infty(Q_T)} = O(\epsilon^{5/8}),$$

$$\|\overrightarrow{u}_{\epsilon x x}(\cdot, t) - \overrightarrow{u}_{xx}(\cdot, t)\|_{L_\infty(Q_T)} = O(\epsilon^{1/8}),$$

$$\|\overrightarrow{u}_{\epsilon x x x}(\cdot, t) - \overrightarrow{u}_{xxx}(\cdot, t)\|_{L_\infty(Q_T)} = O(\epsilon^{1/8}).$$

Let $D \to \infty$, we have

Theorem 3.4 Suppose that the following conditions are satisfied:

(1) $\overrightarrow{u}_0(x) \in W_2^{(2)}(-\infty, \infty)$,

(2) $\overrightarrow{f}(x, t, \overrightarrow{u}) \in C^{(2,1,2)}$, $\overrightarrow{f}(x, t, 0) = 0$. Furthermore, there is the estimate

$$|\overrightarrow{f}_x(x, t, \overrightarrow{u})| \leq c(x, t)|\overrightarrow{u}|^3 + d(x, t),$$

where $c(x, t)$ is bounded in $Q_T^* = \{x \in R, 0 \leq t \leq T\}$, $d(x, t) \in L_2(Q_T^*)$,

(3) the Jacobi derivative matrix $\overrightarrow{f}_u(x, t, \overrightarrow{u})$ is semi-bounded,

(4) $\phi(\overrightarrow{u}) \in C^4$, and conditions ($\Phi$).

Then there exists a unique global weak solution $\overrightarrow{u}(x,t)$ of problem (3.4)(3.6), and

$$\overrightarrow{u}(x,t) \in L_\infty(0,T; W_2^{(2)}(-\infty,\infty)) \cap W_\infty^{(1)}(0,T; W_2^{(1)}(-\infty,\infty)).$$

For the system of generalized KdV type equation with higher order [32]

$$\overrightarrow{u}_t + (\mathrm{grad}\varphi(\overrightarrow{u}))_x + \overrightarrow{u}_{x^{2p+1}} = \overrightarrow{f}(x,t,\overrightarrow{u}), \quad p \geq 1 \tag{3.8}$$

we can also consider the periodic initial value problem and Cauchy problem, and obtain the similar results.

There are some generalizations of Korteweg-de Vries equations which contain nonlinear terms with derivatives of higher order. these can be regarded as special cases of the following fairly wide class of general systems of Korteweg-de Vries type

$$\overrightarrow{u}_t + \overrightarrow{u}_{x^{2p+1}} + G(\overrightarrow{u}, \overrightarrow{u}_x, \cdots, \overrightarrow{u}_{x^{2s}})_x = A(x,t)\overrightarrow{u} + \overrightarrow{g}(x,t), \tag{3.9}$$

where the nonlinear part $G(u, u_x, \cdots, u_{x^{2s}})$ takes the special form

$$G(u, u_x, \cdots, u_{x^{2s}}) = \sum_{k=0}^{s}(-1)^k(\mathrm{grad}_k F(u, u_x, \cdots, u_{x^s}))_{x^k}, \tag{3.10}$$

$\overrightarrow{u} = (u_1, \cdots, u_N)$ is the N-dimensional vector valued unknown function, $A(x,t)$ is an $N \times N$ matrix of functions, $\overrightarrow{g}(x,t)$ is an N- dimensional vector valued function, $F(p_0, p_1, \cdots, p_k)$ is a scalar functions of $(s+1)$ vector variables $p_0, p_1, \cdots, p_s \in R^N$ of dimension N, "grad_k" $(k = 0, 1, \cdots, s)$ denotes the gradient operator with respect to the vector variables $p_k = (p_{k1}, \cdots, p_{kN})$ $(k = 0, 1, \cdots, s)$ and $p \geq 1$ is a natural number.

By using Leray-Schauder fixed point argument, we can prove that

Theorem 3.5 Suppose that the following conditions are fulfilled:

(1) The $(s+2)$-times continuously differentiable scalar function $F(p_0, \cdots, p_s)$ of $(s+1)$ vector variables $p_0, \cdots, p_s \in R^N$ of dimension N satisfies Condition (F)

(F): For the partial derivatives of F, there is

$$|F_{p_{k_1 i_1}, p_{k_2 i_2}, \cdots, p_{k_r i_r}}(p_0, \cdots, p_s)| \leq K_1 (\sum_{j=1}^{s}|p_j|^{\frac{1}{2j+1}})^{l-\sum_{j=0}^{r}(2k_j+1)}, \tag{3.11}$$

where $l = 4p + 2 - \delta$, $\delta > 0$; $k_1, \cdots, k_r = 0, 1, \cdots, s$; $i_1, \cdots, i_r = 1, 2, \cdots, N$; $0 \leq r \leq s+2$, and K_1 is a constant,

(2) The N-dimensional vector function $\varphi(x) \in H^{p+1}(R)$,

(3) The $N \times N$ matrix $A(x,t) \in L_\infty(0,T; W^{(p)}(R))$, and $(A\xi, \xi) \leq b|\xi|^2$, $\xi \in R^N$, b =const, and the N-dimensional vector valued function $\overrightarrow{g}(x,t)$ with components belonging to $L_2(0,T; H^p(R))$, $\overrightarrow{g}(x,t) \in L_\infty(Q_\infty)$, $Q_\infty = R \times [0,T]$. Then the initial value problem (3.12) for the general system (3.9) (3.10) of Korteweg-de Vries type has at least one global N-dimensional vector valued weak solution $\overrightarrow{u}(x,t) \in L_\infty(0,T; H^p(R)) \cap C^{(0, \frac{s}{2p+2})}(0,T; L_2(R))$. and

$$\overrightarrow{u}(0,x) = \varphi(x), \quad x \in R. \tag{3.12}$$

Now let us give a remark on the "blowing up" behavior of N-dimensional vector valued generalized solution of the initial value problem for the general system

$$\overrightarrow{u}_t + \overrightarrow{u}_{x^{2p+1}} + G(\overrightarrow{u}, \overrightarrow{u}_x, \cdots, \overrightarrow{u}_{x^{2s}})_x = \overrightarrow{f}(x, t, \overrightarrow{u}) \tag{3.13}$$

of Korteweg-de Vries type. Suppose that the N-dimensional vector valued right hand part of the system (3.13) has the property

$$\int_{-\infty}^{\infty} (\overrightarrow{u}, \overrightarrow{f}(x, t, \overrightarrow{u})) dx \geq C_0 \|\overrightarrow{u}(\cdot, t)\|_{L_2(R)}^2 \psi(\|\overrightarrow{u}(\cdot, t)\|_{L_2(R)}), \tag{3.14}$$

where $C_0 > 0$ and $\psi(z)$ satisfies

$$\int_0^{\infty} \frac{dz}{z\psi(z)} < \infty. \tag{3.15}$$

Taking scalar product of system (3.13) with vector $\overrightarrow{u}$ and integrating the resulting relation for x in R, we have

$$\frac{d}{dt} \|\overrightarrow{u}(\cdot, t)\|_{L_2(R)}^2 = 2 \int_{-\infty}^{\infty} (\overrightarrow{u}, \overrightarrow{f}(x, t, \overrightarrow{u})) dx.$$

This becomes

$$\frac{d}{dt} w(t) \geq C_0 w(t) \psi(w(t)),$$

where $w(t) = \|\overrightarrow{u}(\cdot, t)\|_{L_2(R)}$. Hence $w(t)$ goes to infinity at a finite value of t.

Theorem 3.6 Suppose that $\overrightarrow{f}(x, t, \overrightarrow{u})$ satisfies conditions (3.14) and (3.15). Then the N-dimensional vector valued generalized solution of the initial value problem (3.12) for the general system (3.13) of Korteweg-de Vries type has an infinite norm $\|\overrightarrow{u}(\cdot, t)\|_{L_2(R)}$ at certain finite time t.

We also consider the following nonlinear 5th order equation of KdV type [33]

$$u_t + \left(\frac{\partial F(u)}{\partial u}\right)_x + \left(\frac{\partial G(u, u_x)}{\partial u}\right)_x - \left(\frac{\partial G(u, u_x)}{\partial u_x}\right)_{xx} + u_{x^5} = 0, \tag{3.16}$$

with the initial condition

$$u|_{t=0} = \varphi(x), \tag{3.17}$$

with $G(u, u_x) = a_1 u u_x^2 + a_2 u^2 u_x^2 + a_3 u^3 u_x^2 + a_4 u_x^3$, a_i $(i = 1, 2, 3, 4)$ are constants. It is obvious which includes the 5th KdV equation

$$u_t + 30u^2 u_x + 10(2u_x u_{xx} + u u_{xxx}) + u_{x^5} = 0, \tag{3.18}$$

founding by P.D. Lax, and for the equation (3.18), the infinitely many conservation laws and inverse scattering form have been obtained.

By using the vanishing viscosity method and a priori estimates, we can prove

Theorem 3.7 Suppose that the following conditions are satisfied:
(1) $F(u) \in C^{s+2}(R)$, $F'(0) = F(0) = 0$, and

$$|F''(u)| \leq A_1(1 + |u|^7), \quad A_1 = const > 0 \tag{3.19}$$

(2) $\varphi(x) \in H^s(R)$, $s \geq 5$.
Then the initial value problem (3.16) (3.17) exists at least one smooth solution

$$u(x,t) \in \cap_{k=0}^{[s/5]} W_\infty^k(0,T;H^{s-5k}(R)) \cap (\cap_{k=0}^{[(s+1)/5]} W_\infty^k(0,T;H_{loc}^{s+1-5k}(R))).$$

Now we consider systems of the generalized Korteweg-de Vries type of higher order with several variables of the form [34]

$$u_t + \delta\delta_{2p}\vec{u} + \delta(\mathrm{grad}\varphi(\vec{u})) = A(x,t)\vec{u} + \vec{g}(x,t), \qquad (3.20)$$

where $\vec{u}(x,t) = (u_1(x,t),\cdots,u_N(x,t))$ denotes an unknown function of variables $x = (x_1,\cdots,x_n) \in R^n$, and $t \in R^+$, $\varphi(\vec{u}) = \varphi(u_1,\cdots,u_N)$ means a scalar function of vector variables $\vec{u} \in R^N$. $\vec{g}(x,t)$ is a function of N-dimensional vector value and $A(x,t)$ refers to an $N \times N$ matrix of functions. Hence $\delta_k = \sum_{i=1}^n D_i^k$ is a differential operator of k order, where $D_i = \frac{\partial}{\partial x_i}$ $(i = 1,\cdots,n;\ k = 1,2,\cdots)$, and "grad" denotes the gradient operator with respect to the vector variable $\vec{u}$. $\vec{u}_0(x)$ is a given function of N-dimensional initial vector.

By using Leray-Schauder fixed point argument, we can prove that

Theorem 3.8 Suppose that the following conditions are satisfied:
(1) The function of N-dimensional vector value $\vec{u}_0(x) \in H^{p+1}(R^n)$,
(2) The $N \times N$ matrix $A(x,t) \in L_\infty(0,T;W_\infty^{(p)}(R^n))$ and the function of N-dimensional vector value $\vec{g}(x,t) \in L_2(0,T;H^p(R^n))$.
(3) The scalar function $\varphi(\vec{u}) \in C^2(R^N)$ satisfies the conditions
(i) $|\varphi(\vec{u})| \leq K|\vec{u}|^{s+2}$,
(ii) $|\mathrm{grad}\varphi(\vec{u})| \leq K|\vec{u}|^{s+1}$,
(iii) $|\frac{\partial^2\varphi(\vec{u})}{\partial u_i\partial u_j}| \leq K(|\vec{u}|^s + 1)$, $(i,j,= 1,2,\cdots,N)$, where $s = 4p/n - \delta_1 > 0$, $\delta_1 > 0$ and K is a constant. Then the initial value problem (3.21) for the systen (3.20) of generalized Korteweg-de Vries type of higher order $(p \geq 1)$ with several variables has at least one global weak solution

$$\vec{u}(x,t) \in L_\infty(0,T;H^p(R^n)) \cap C^{(0,\frac{p}{2p+2})}(0,T;L_2(R^n)) \qquad (3.21)$$

of N-dimensional vector value.

Theorem 3.9 Suppose that the conditions (1)(2)(3) of Theorem 3.8 are satisfied, and assume that
(4) $2p > n$, and $\varphi(\vec{u}) \in C^3(R^N)$.
Then the generalized solution $\vec{u}(x,t)$ for the initial value problem (3.20) (3.21) is unique.

Finally, we consider the following of coupled KdV-nonlinear Schrodinger equations and KdV-nonlinear wave equations [35][36]

$$i\vec{\epsilon}_t + \alpha\vec{\epsilon}_{xx} - \beta n\vec{\epsilon} = 0, \qquad (3.22)$$

$$n_t + 1/2 f(n)_x + \gamma/2 n_{xxx} + 1/2|\vec{\epsilon}|_x^2 = 0, \qquad (3.23)$$

and

$$u_t = u_{xxx} + 6uu_x + 2vv_x, \qquad (3.24)$$

$$v_t = 2(uv)_x,$$ (3.25)

with the initial conditions

$$\vec{\epsilon}\,|_{t=0} = \vec{\epsilon}_0(x), \quad n|_{t=0} = n_0(x), \quad x \in R$$ (3.26)

and

$$u|_{t=0} = u_0(x), \quad v|_{t=0} = v_0(x), \quad x \in R$$ (3.27)

respectively, where $\vec{\epsilon}(x,t) = (\epsilon_1(x,t), \cdots, \epsilon_N(x,t))$ is a complex valued functional vector, $n(x,t)$ is a real function, α, β and γ are real constants.

Theorem 3.10 Suppose that the following conditions are satisfied:
(1) $f(n) \in C^s$, $(s \geq 3)$, and $|f^{(l)}(n)| \leq A|n|^{2-l}$, $l = 0, 1$, $A > 0$.
(2) $\alpha\beta\gamma < 0$,
(3) $\vec{\epsilon}_0(x), n_0(x) \in H^s(R)$, $(s \geq 3)$.
Then there exists a unique global solution for the initial value problem (3.22)(3.23) (3.26), and

$$\vec{\epsilon}(x,t) \in L_\infty(0,T; H^s(R)), \quad n(x,t) \in L_\infty(0,T; H^s(R)).$$

Theorem 3.11 Let T be any given positive constant. For any initial data $u_0(x), v_0(x)$ such that $(u_0(x), v_0(x)) \in H^k(R)$, where $k \geq 4$ is an integer number, then the initial value problem for the KdV-nonlinear wave equations (3.24)(3.25)(3.27) has a unique smooth solution (u, v), and

$$(u, v) \in \cap_{s+3r \leq k} W_\infty^r(0,T; H^s(R)).$$

References

[1] M. Lakshmanan and K. Nakamura, Landau-Lifshitz equation of ferromagnetism : Exact Treatment of the Gilbert Damping, Phy. Rew. Lett., 53(26), 1984, 24976-2499.

[2] L.D. Landau and E.M. Lifshitz, On the theory of the dispersion of magnetic permeability in ferromagnetic bodies, Phys. Z. Sowej., 8, 1935, 153. Reproduced in 1965, 101-114, Collected paper of L.D. Landau (D. ter Hear, eds) , Pergamon, New York.

[3] H.C. Fogedby, Theoretical aspects of mainly low dimensional magnetic systems, Lecture Notes in Phys. 131, Springer-Verlag, Berlin, Heideberg, 1980.

[4] K. Nakamura and T. Sasada, Soliton and wave trains in ferromagnets, Phys, Lett., 48(A), 1974, 321-322.

[5] L.A. Takhtalian, Integration of the continuous Heisenberg spin chain through the inverse scattering method, Phys. Lett. 15(B), 1977, 3470- 3476.

[6] J. Tjon and J. Wright, Soliton in the continuous Heisenberg chain, Phys. Rew.

[7] Zhou Yulin and Guo Boling, On the solvability of the initial value problem for the quasilinear degenerate parabolic system $\vec{Z}_t = \vec{Z} \times \vec{Z}_{xx} + \vec{f}(x, t, \vec{Z})$, Proceedings of DD-3 Symposium, 1982, 713-732.

[8] Zhou Yulin and Guo Boling, Existence of weak solution for boundary problems of ferromagnetic chain, Scientia Sinica (A), 27(8), 1984, 799-811.

[9] Zhou Yulin and Guo Boling, Some boundary problems of the spin system and the system of ferromagnetic chain I: Nonlinear boundary problems, Acta Math. Scientia, 6(3), 1986, 321-337.

[10] Zhou Yulin and Guo Boling, Some boundary problems of the spin system and the system of ferromagnetic chain II: Mixed problems and others, Acta Math. Scientia, 7(2), 1987, 121-132.

[11] Zhou Yulin and Guo Boling, The weak solution of homogeneous boundary value problem for the system of ferromagnetic chain with several variables, Scientia Sinica, 4(A)(1986), 337-349.

[12] Boling Guo and Min-Chun Hong, The Landau-Lifshitz equation of the ferromagnetic spin chain and harmonic maps, to appear in "Calculus of Variations and PDE".

[13] Zhou Yulin, Sun Hesheng and Guo Boling, Geometrical extensions for systems of ferromagnetic chain, Advances in Math., China, 21(4), 1992, 497-501.

[14] Zhou Yulin, Guo Boling and Tan Shaobin, Existence and uniqueness of smooth solution for system of ferromagnetic chain, Science in China, 34(A), 1991, 257-266.

[15] J. Satsuma and D.J. Kaup, A Backlund transformation for a higher order Korteweg-der Vries equation, J. Phys. Soc. Japan, 43, 1977, 692- 697.

[16] J. Satsuma and R. Hirota, A coupled KdV equation in one case of four- reduction of the KP Hierarchy, J. Phys. Soc. Japan, 51, 1982, 3390-3397.

[17] R. Hirota and M. Ito, Resonance of solitons in one dimension, J. Phys. Soc. Japan, 52, 1983, 744-748.

[18] Zhou Yulin and Guo Boling, Initial value problems for a nonlinear singular integral-differential equation of deep water, Proceedings of DD-7 Symposium, held in Tianjin, June, 23-July 5, 1986, Lecture Notes in Math., 1306, 278-290.

[19] Rafael Jose Iorio, Jr., On the Cauchy problem for the Benjamin- Ono equation, Comm. in Partial Differential Equations, 11, 1986, 1031-1084.

[20] Zhou Yulin and Guo Boling, Global solutions and their large time behavior of Cauchy problem for equations of Benjamin-Ono type, to appear.

[21] Zhou Yulin, Guo Boling and Tan Shaobin, On the Cauchy problem for the equation of finite-depth fluids, J. Partial Differential Equations, 5(1), 1992, 1-16.

[22] Guo Boling and Tan Shaobin, Cauchy problem for a generalized nonlinear dispersion equations, J. Partial Differential Equations, 5(4), 1992, 37-50.

[23] E. Ott and R.N. Sudan, Phys. Fluid, 12, 1969, 2388-2394.

[24] A. Sjoberg, On the Korteweg-der Vries equation: Existence and uniqueness, J. Math. Anal. Appl., 29, 1970, 569-579.

[25] J.L. Bona and R. Smith, The initial-value problem for the Korteweg-der Vries equation, Phil. Trans. Roy. Soc. London, 278(A), 1975, 555-601.

[26] J.C. Saut, Sur quelques generalisations de lequation de Korteweg-der Vries, J. Math. pure appl., 58, 1979, 21-61.

[27] J. Ginibre, Y. Tsutsumi and G. Velo, Existence and uniqueness of solutions of the generalized Korteweg-der Vries equation,???

[28] Charies F.F. Karney, Abhijit Sen, Flora Y.F. Chu, Nonlinear evlution of lower hybrid wave, Phys. Fluid, 22(5), 1979, 940-956.

[29] Guo Boling, The global solution for one class of generalized KdV equation, Acta Math. Sinica, 25(6), 1982, 641-656.

[30] Zhou Yulin and Guo Boling, On the system of the generalized Korteweg-der Vries equations, Proceedings of DD-3 Symposium, 1982, 739-758.

[31] Zhou Yulin and Guo Boling, The periodic boundary problems and the initial value problems for the system of generalized Korteweg-der Vries type of high order, Acta Math. Sinica, 27(2), 1984, 154-176.

[32] Zhou Yulin and Guo Boling, A class of general systems of Korteweg-der Vries type I: Weak solutions with derivative, Acta Math. Appl. Sinica, 2,1984.

[33] Guo Boling, Han Yongqian and Zhou Yulin, On smooth solution for a nonlinear 5th order equation of KdV type, to appear.

[34] Zhou Yulin and Guo Boling, Existence of the global weak solutions for the systems of generalized KdV type of higher order with several variables, Scientia Sinica, 29(4), 1986, 375-390.

[35] Guo Boling, Existence and uniqueness of the global solution of the Cauchy problem for a class of the coupled system of KdV-nonlinear Schrodinger equations, Acta Math. Sinica, 26(5), 1983, 513-532.

[36] Guo Boling and Tan Shaobin, Global smooth solution for a coupled nonlinear wave equations, Math. Meth. Appl. Sci., 14(6), 1991, 419-425.

Solitons and Exactly Solvable Nonlinear Evolution Equations†

Yi Cheng and Yi-shen Li

Department of Mathematics University of Science and Technology of China, Hefei

1 Introduction

The soliton is an analytic and exact solution to classes of nonlinear evolution equations. It was discovered in 1965 by Zabusky and Kruskal [1], who were experimenting with numerical solution by computer of the Korteweg-de Vries (KdV) equation. This is a nonlinear partial differential equation and had been introduced at the last century to describe wave motion in shallow canals. Zabusky and Kruskal studied it because of its relevance to the singnificance in theoretical physics, and named " soliton" to describe the solitary wave solutions of the KdV because of their partical like behaviour. Soon afterwards, Gardner, Greene, Kruskal and Miura (GGKM) [2] discovered the so-called inverse scattering transform (IST) method to find exact general solutions to the KdV equation. Since then the soliton theory has been advancing rapidly and the IST as well as other methods in the theory have become most powerful tools for solving large class of nonlinear evolution equations (NEEs) including some of physically interesting ones, (usually, the NEEs possessing soliton solutions are called soliton equations). Some other related topics have also been arised, which cover many areas in both mathematics and physics.

Starting in 1976 (the year of the end of " Cultural Revolution"), many Chinese mathematicians and theoretical physicists began to work in the field of solitons and exactly solvable nonlinear equations. They mainly include, from mathematical aspect, Cao Ce-wen in Zhengzhou, Gu Chao-hao and Hu He-shen in Shanghai, Qian Ming, Tu Gui-zhang and Zhuang Da-wei in Beijing and the group leading by Li Yi-shen in Hefei, as well as many others.

In the present article we would like to survey our works carried out by us and our collaborators during eighties on solitons and exactly solvable NEEs. In the next section, we give some results on the extension of IST to non-isospectral flows of soliton equations in 1+1 (i.e., one spatial and one temporal) and in 2+1 (i.e., two spatial and one temporal) dimensions (the geometrical meaning of non-isospectral

† Supported by National Basic Research Project for "Nonlinear Science"
† Supported by NNSF of China

C. Gu et al. (eds.), Partial Differential Equations in China, 160–172.
© 1994 Kluwer Academic Publishers.

flows will be given in Section 5). We will also show an investigation on the discrete AKNS system.

In Section 3, we consider the gauge equivalence (GE) between different spectral problem and between the correspondent soliton equations. As a special case that the gauge transformations is performed for the same spectral problem and the same equation, the GE leads to the Backlund and Darboux transformations which transform the solution to another solution of the same equation under consideration. These transformations are known to be other powerful methods for solving the soliton equations. We will discuss them in Section 4.

It is known that most of soliton equations have the Hamiltonian structures which are in the infinite dimensions. The groups of symmetry transformations, their related algebraic property and the existence of conserved quantities turn out to be important. In Section 5, we study these features, in particular the construction of symmetries for soliton equations in both 1+1 and 2+1 dimensions. the group of symmetry transformations corresponding to symmetries are obtained by solving the initial value problem of isospectral and non-isospectral flows.

2   On inverse scattering transform

2.1   Non-isospectral flows

The IST method proposed by GGKM [2] for solving the KdV equation has been extended to many other NEEs. The crucial steps for the applications of IST to a given NEE, say, $u_t(x,t) = K(u(x,t))$, are as follows.

1) Finding a spectral problem

$$L\psi = \lambda\psi \qquad (2.1)$$

which contains u(x,t) as potential and can be performed the direct and inverse scattering transforms.

2) Finding an Lax operator A, such that the time evolution for ψ is given by

$$\psi_t = A\psi \qquad (2.2)$$

and such that the compatibility condition of (2.1) and (2.2) under the condition of $\lambda_t = 0$ (i.e., the isospectral) is equivalent to the given equation. This compatibility condition with $\lambda_t = 0$ is written as the following Lax representation

$$L_t = [A, L] \qquad (2.3)$$

Conversely, any equation obtained from the Lax representation (2.3) is solvable in principle via the IST of the spectral problem (2.1).

If we require that

$$\lambda_t = \sum_{k=0}^{m} \varepsilon_k \lambda^k \tag{2.4}$$

the Lax representation obtained by the compatibility condition of (2.1) and (2.2) is extended to

$$L_t = [A, L] + \sum_{k=0}^{m} \varepsilon_k L^k \tag{2.5}$$

from which we also obtain a class of NEEs solvable via the non-isospectral deformations of IST.

Let us take the AKNS spectral problem

$$L\psi = i\lambda\psi, \qquad L = -\sigma_3 \partial_x + u(x, t) \tag{2.6}$$

as the example, where $u(x, t) = q(x, t)\sigma_+ - r(x, t)\sigma_-$, $\sigma_\pm = \sigma_1 \pm i\sigma_2$ and $\sigma_1, \sigma_2, \sigma_3$ are the Pauli matrices. The potential u(x,t) in (2.6) is always assumed to be decay rapidly enough as $x \to \infty$. To derive the NEEs associated with (2.6), we look for the Lax operator

$$A(u) = \sum_{j=0}^{m} A_j(u) L^j$$

being the polynomial in L such that the right hand side of (2.5) be the off-diagonal and multiplication matrix. It is noted that [A, L] can also be expressed as polynomial in L (see [3]). and so equating the coefficients of the powers L^j, $(j > 0)$ in right hand side of (2.5) to be zero, the coefficients of A can be derived recursively by integrations. The particular choices of the integral constants and ε_j, j=1,...,m give rise to two hierarchies of Lax operators A_n and B_n such that

$$[A_n, L] = a_n(u) = a_n^{(1)}(u)\sigma_+ - a_n^{(2)}(u)\sigma_- \tag{2.7}$$
$$[B_n, L] + L^n = b_n(u) = b_n^{(1)}(u)\sigma_+ - b_n^{(2)}(u)\sigma_- \tag{2.8}$$

where $a_n^{(i)}(u)$ and $b_n^{(i)}$, i=1,2 are multiplications. The correspondent isospectral $(\lambda_t = 0)$ and non-isospectral $(i\lambda_t = (i\lambda)^n)$ flows are then given by

$$u_t = a_n(u) \tag{2.9}$$

and

$$u_t = b_n(u) \tag{2.10}$$

respectively. Equations in (2.9) are members of the well-known AKNS hierarchy [4] which includes the generalized nonlinear Schrodinger (NS) equation

$$q_t = q_{xx} + 2q^2 r$$
$$r_t = -r_{xx} - 2qr^2 \tag{2.11}$$

as a typical example. Both (2.9) and (2.10) can be solved by IST and the correspondent time evolutions of the scattering date $\{(\lambda_j, \beta_j)_{j=1}^N, (\bar{\lambda}_j, \bar{\beta}_j)_{j=1}^N, \rho(\lambda), \bar{\rho}(\lambda), Im\lambda = 0\}$ of (2.6) necessary to carry out the IST are given by

$$\lambda_t = 0, \quad \rho_t(\lambda) = -i\lambda^n \rho(\lambda)$$
$$\lambda_{j,t} = 0, \quad \beta_{j,t} = -i\lambda_j^n \beta_j \tag{2.12a}$$

for (2.9) and

$$\lambda_t = \lambda^n, \quad \rho_t(\lambda) = \frac{\partial \rho}{\partial t} + \frac{\partial \rho}{\partial \lambda}\lambda^n = 0$$
$$\lambda_{j,t} = \lambda_j^n, \quad \beta_{j,t} = n\lambda_j^{n-1}\beta_j \tag{2.12b}$$

for (2.10) and similar equations for $\overline{\lambda}, \overline{\beta}$ etc. The linear combination of (2.9) and (2.10) are also solvable via the IST, which include

$$q_t = q_{xx} + 2q^2 r + xq$$
$$r_t = -r_{xx} - 2qr^2 + xr \tag{2.13}$$

as a simple example. This equation can be used to describe the solitary wave travelling in an inhomogenous medium [5].

The results for non-isospectral flows shown above are from [6], while the derivations of the Lax operators are from [3] because of the convenience for late uses.

There are many other studies dealing with the non-isospectral deformations of IST, such as the discussion for the Jaulent-Miodek spectral problem

$$\varphi_{xx} + [\lambda^2 - (u(x) + \lambda v(x)]\varphi = 0$$

and

$$\varphi_{xx} + [x + u(x)]\varphi = \lambda\varphi$$

which is used for solving the symmetric KdV equations (see [7, 8] for detail).

The above discussion has been generalized to the 2+1 dimensional soliton equations by Cheng, Li and Bullough; and Cheng [9-12]. These equations include the Kadomtsev-Petviashvili (KP) equation, the modified KP, the Kotera-Sawada equations as well as the integro-differential Benjamin-Ono equation.

As for the KP equation

$$u_t = K_2(u) = \partial_x^{-1} u_{yy} - u_{xxx} - 6uu_x \tag{2.14}$$

where u=u(x,y,t), the spectral problem is the following "non-stationary" Schrödinger equation

$$L\psi = 0, \quad L = \alpha\partial_y + \partial_x^2 + u, \quad \alpha = \frac{i}{3}\sqrt{3} \tag{2.15}$$

Assicisted with (2.15), there are two hierarchies of equations, one of them is isospectral

$$u_t = K_n(u), \quad n \geq 0 \tag{2.16}$$

including the KP equation (2.14) (n=2) as an example and the other is non-isospectral

$$u_t = \sigma_n(u), \quad n \geq 1 \tag{2.17}$$

Both (2.16) and (2.17) have the Lax representations

$$L_t = K_n(u) = [A_n, L] \tag{2.18}$$

$$L_t = \sigma_n(u) = [B_n, L] \tag{2.19}$$

The Lax operators A_n and B_n are $(n+1)$th and nth order of polynomials in ∂_x with leading term coefficients being constant and proportional to y respectively. Unlike in the case of 1+1 dimensions, the Lax operators A_n and B_n for the KP are derived by the following algebraically recursive formulae

$$A_{n+1} = \frac{3}{n+1}||[A_n, B_3]||, \quad n \geq 0 \tag{2.20}$$

$$B_{n+1} = \frac{3}{n-3}||[B_n, B_3]||, \quad n \geq 4 \tag{2.21}$$

where the bracket for any two Lax operators A and $\overline{A}$ is defined as

$$||[A, \overline{A}]|| = A'[\overline{a}] - \overline{A}'[a] + [A, \overline{A}]$$
$$A'[\overline{a}] = \lim_{\varepsilon \to 0} \frac{\partial}{\partial \varepsilon} A(u + \varepsilon \overline{a}), \quad etc. \tag{2.22}$$

and a=[A, L], $\overline{a} = [\overline{A}, L]$ are multiplication operators. The first few Lax operators can be calculated directly from the Lax representation, they are

$$\begin{aligned}
&A_0 = \tfrac{1}{3}\partial_x, \quad A_1 = 2\alpha(\partial_x^2 + u) \\
&A_2 = -4\partial_x^3 - 6u\partial_x - 3u_x + i\sqrt{3}(\partial_x^{-1}u_y) \\
&B_1 = \tfrac{1}{3}yA_0 - \tfrac{1}{2}\alpha x, \quad B_2 = yA_1 + xA_0 + \tfrac{1}{3} \\
&B_3 = yA_2 + xA_1 + 2\alpha\partial_x + \alpha(\partial_x^{-1}u) \\
&etc.
\end{aligned} \tag{2.23}$$

(The operator B_4 can also be derived but is more complicated, its exact form was displayed in [9,12]). The time evolutions of the scattering data of (2.15) can be derived such that (2.16) and (2.17) are solvable via IST of (2.15), and the spectra for (2.16) are time independent ($k_t = 0$) and for (2.17) are time dependent ($k_t = -\frac{i}{6\sqrt{3}}(-2\sqrt{3}k)^{n-1}$). Detail can be found in [9, 12].

2.2   Equivalence of IST of two discrete spectral problems

In [13], Ablowitz and Ladik proposed two discrete versions of AKNS spectral problems, they are

$$\psi_{n+1} = \begin{pmatrix} z & \overline{Q}_n \\ \overline{R}_n & z^{-1} \end{pmatrix} \psi_n \tag{2.24}$$

and

$$\varphi_{n+1} = \begin{pmatrix} z & Q_n \\ R_n & z^{-1} \end{pmatrix} \varphi_n + \begin{pmatrix} 0 & S_n \\ T_n & 0 \end{pmatrix} \varphi_{n+1} \tag{2.25}$$

where z is the spectral parameter and n plays the role as the discrete variable. Both of them can be performed the discrete version of IST and both of them are associated with a class of nonlinear differential-difference equations (NDDEs) solvable via the IST. These NDDEs include a discrete version of NS equation, the net-work lattice and Toda lattice [13].

At first glance, (2.25) is more general than (2.24) because it contains four potentials, and the reduction $S_n = T_n = 0$ leads (2.25) to (2.24). However, as was shown by Cheng in [14], if we first let

$$\chi_n(z) = \prod_{j=n}^{+\infty} \Lambda_j^{-1} \rho(z) \varphi_n(z^2),$$
(2.26)

$$\Lambda_j = 1 - S_j T_j, \quad \rho(z) = \begin{pmatrix} z^{\frac{1}{2}} & \\ & z^{-\frac{1}{2}} \end{pmatrix}$$

χ_n satisfies

$$\chi_{n+1} = \begin{pmatrix} z & S_n \\ T_n & z^{-1} \end{pmatrix} \begin{pmatrix} z & Q_n \\ R_n & z^{-1} \end{pmatrix} \chi_n$$
(2.27)

then we define

$$\psi_{2n}(z) = \chi_n(z), \quad \psi_{2n+1}(z) = \begin{pmatrix} z & Q_n \\ R_n & z^{-1} \end{pmatrix} \chi_n(z)$$
(2.28)

and potantials

$$\overline{Q}_n = \{ Q_k, \quad n = 2kS_k, \quad n = 2k+1, \qquad \overline{R}_n = \{ R_k, \quad n = 2kT_k, \quad n = 2k+1$$
(2.29)

one can easily check that ψ_n in (2.28) satifies (2.24) with potentials given by (2.29). The scattering data, the procedure of IST and the NDDEs associated with (2.24) and (2.25) respectively are identified by the correspondences (2.28) and (2.29). We conclude, therefore, that the spectral problem (2.25) is not at all more general than (2.24) (see [14] for further information).

3 Gauge equivalences among soliton equations

Since the discovery of solitons and IST method, many nonlinear evolution equations solvable via the IST have been found. These equations are either from physics or from the Lax representation for some new spectral problems and new Lax operators. We shall show in this section, some of them are gauge equivalent to each other. The famous example of the gauge equivalence (GE) was given by Zakharov and Takhtajan [15] which takes place between the NS equation

$$iq_t = -q_{xx} + 2c|q|^2 q$$
(3.1)

and the Heisenberg magnet (HM) equation

$$S_t = [S, S_{xx}]$$
(3.2)

where $S = s_1\sigma_1 + s_2\sigma_2 + s_3\sigma_3$, $S^2 = I$.

To explain the meaning of GE, let us consider two linear systems

$$\varphi_x = M(u, \lambda)\varphi, \quad \varphi_t = N(u, \lambda)\varphi \tag{3.3}$$

and

$$\overline{\varphi}_x = \overline{M}(\overline{u}, \overline{\lambda})\overline{\varphi}, \quad \overline{\varphi}_t = \overline{N}(\overline{u}, \overline{\lambda})\overline{\varphi} \tag{3.4}$$

The first equations in each of (3.3) and (3.4) are considered as the spectral problem containing their own potentials. The compability condition $\varphi_{xt} = \varphi_{tx}$ gives rise to the Lax representation (or zero courture condition)

$$M_t - N_x = [N, M] \tag{3.5}$$

and $\overline{\varphi}_{xt} = \overline{\varphi}_{tx}$ leads to the same equation for $\overline{M}$ and $\overline{N}$. From (3.5) and equation for $\overline{M}$ and $\overline{N}$, classes of NEEs corresponding to (3.3) and (3.4) respectively are obtained.

If there exists a gauge transformation

$$\overline{\varphi} = T\varphi \tag{3.6}$$

connecting the eigenfunctions of both (3.3) and (3.4), then the gauge matrix T must satisfy

$$T_x = \overline{M}T - TM \tag{3.7}$$

$$T_t = \overline{N}T - TN \tag{3.8}$$

Conversely, if T is solved from (3.7) and (3.8), the compability condition $T_{xt} = T_{tx}$ leads

$$(\overline{M}_t - \overline{N}_x + [\overline{M}, \overline{N}])T = T(M_t - N_x + [M, N])$$

which implies that if one pair (M, N) or $(\overline{M}, \overline{N})$ satisfies the Lax representation (3.5), then so does the other. In this sense we say that the NEEs obtained from (3.5) for (M, N) or $(\overline{M}, \overline{N})$ respectively are gauge equivalence (GE). In [16, 17], Li and his coworkers constructed the GE between the AKNS eigenvalue problem

$$\varphi_x = M\varphi, \quad M = -i\lambda\sigma_3 + q\sigma_+ + r\sigma_- \tag{3.9}$$

and the Kaup-Newell one, and between the correspondent equations. Soon after they [18] generalized the GE of (3.9) and a more general eigenvalue problem

$$\overline{\varphi}_x = \overline{M}\overline{\varphi}, \atop \overline{M} = -[(i\lambda\omega_3 - p_3)\sigma_3 + (i\lambda\omega_1 - p_1)\sigma_+ + (i\lambda\omega_2 - p_2)\sigma_-] \tag{3.10}$$

The gauge matrix T between (3.9) and (3.10) is given by

$$T = \frac{1}{\sqrt{2(1+\omega_3)}} \begin{pmatrix} \beta(1+\omega_3) & \beta\omega_1 \\ -\beta^{-1}\omega_2 & \beta^{-1}(1+\omega_3) \end{pmatrix} \tag{3.11}$$

$$\beta = exp[-\int (p_3\omega_3 + \frac{1}{2}(p_1\omega_2 - p_2\omega_1) + \frac{\omega_1\omega_{2,x} - \omega_2\omega_{1,x}}{4(1+\omega_3)})dx] \tag{3.12}$$

such that

$$q = \beta^2 [-p_3\omega_1 + \frac{p_1}{2}(1+\omega_3) - \frac{p_2\omega_1^2}{2(1+\omega_3)} + \frac{\omega_{1,x}}{2} - \frac{\omega_1\omega_{3,x}}{4(1+\omega_3)}] \tag{3.13a}$$

$$r = \beta^{-2}[-p_3\omega_2 + \frac{p_2}{2}(1+\omega_3) - \frac{p_1\omega_2^2}{2(1+\omega_3)} - \frac{\omega_{2,x}}{2} + \frac{\omega_2\omega_{3,x}}{4(1+\omega_3)}] \tag{3.13b}$$

The spectral problem (3.10) and the associated NEEs admit several reductions which reduce them to some physically interesting systems [18].

In [19], Cheng, Li and Tang gave a GE between two typical soliton equations in 2+1 dimensions. One of them is the Ishimori (the 2+1 dimensional analogue of HM) equation

$$\begin{aligned} &S_t + [S, S_{xx} + \alpha^2 S_{yy}] + u_x S_y + u_y S_x = 0 \\ &u_{xx} - \alpha^2 u_{yy} - i\alpha^2 trS[S_x, S_y] = 0 \end{aligned} \tag{3.14}$$

where $\alpha^2 = \pm 1$, $S = s_1\sigma_1 + s_2\sigma_2 + s_3\sigma_3$ satisfying $S^2 = I$. The other is the general Davey-Stewartson (DS) equation (the 2+1 dimensional analogue of the general NS equation in (2.11))

$$\begin{aligned} &iq_t - q_{xx} - \alpha^2 q_{xx} - 2vq = 0 \\ &ir_t + r_{xx} + \alpha^2 r_{yy} - 2vr = 0 \\ &v_{yy} - \alpha^2 v_{xx} + (qv)_{xx} + \alpha^2(qr)_{yy} = 0, \quad \alpha^2 = \pm \end{aligned} \tag{3.15}$$

Both of them have the Lax pairs (L_1, L_2) and $(\overline{L_1}, \overline{L_2})$, such that (3.14) and (3.15) are equivalent to $[L_1, L_2] = 0$, $[\overline{L_1}, \overline{L_2}] = 0$ respectively. In particular the spectral problems are

$$L_1\psi = (\alpha\partial_y + S\partial_x)\psi = 0 \tag{3.16}$$

for (3.14) and

$$\overline{L_1}\overline{\psi} = (\alpha\partial_y - \sigma_3\partial_x + q\sigma_+ + r\sigma_-)\overline{\psi} = 0 \tag{3.17}$$

for (3.15). We have shown in [19] that a gauge matrix T exists

$$T = diag(\lambda^+, \lambda^-)(S - \sigma_3) \tag{3.18}$$

such that $\overline{L_1}T = TL_1$, $\overline{L_2}T = TL_2$, where $\lambda^\pm$ satisfy

$$\begin{aligned} &\pm 4i(\frac{\lambda^\pm_x}{\lambda^\pm} + \frac{(s_1 \mp is_2)_x(s_1 \pm is_2)}{2(1-s_3)} - \frac{s_{3,x}}{2}) \mp \alpha^3 u_x - u_y = 0 \\ &4i\alpha(\frac{\lambda^\pm_y}{\lambda^\pm} + \frac{(s_1 \mp is_2)_y(s_1 \pm is_2)}{2(1-s_3)} - \frac{s_{3,y}}{2}) \mp \alpha^3 u_x - u_y = 0 \end{aligned} \tag{3.19}$$

which are compatible to each other. By the transformation we have

$$\begin{aligned} q &= \frac{\lambda^+}{\lambda^-}[(\frac{\overline{s}s_{3,x}}{2(1-s_3)} + \frac{\overline{s}_x}{2}) - \alpha(\frac{\overline{s}s_{3,y}}{2(1-s_3)} + \frac{\overline{s}_y}{2})] \\ r &= \frac{\lambda^-}{\lambda^+}[(\frac{ss_{3,x}}{2(1-s_3)} + \frac{s_x}{2}) + \alpha(\frac{ss_{3,y}}{2(1-s_3)} + \frac{s_y}{2})] \end{aligned} \tag{3.20}$$

where $s = s_1 + is_2$, $\overline{s} = s_1 - is_2$. The invertible transformation can also be constructed. Thus we conclude that (3.14) and (3.15) are gauge equivalent to each other. This GE generalizes the result of Zakharov and Takhtajan [15] and is, to our knowledge, the first example of GE in 2+1 dimensions given explicitly.

4   Backlund and Darboux transformations

From the last section, we see that the GE can also be considered as the transformation which transforms the solution of one equation to solution of the other. As we discuss in this section if the gauge transformation takes place for the same equation but between different solutions, we obtain the so-called Backlund and Darboux transformations.

Still take the AKNS hierarchy as the example, we consider

$$\varphi_x = M\varphi, \quad \overline{\varphi}_x = \overline{M}\,\overline{\varphi} \tag{4.1}$$

where M and $\overline{M}$ has the same form as that in (3.9) but contain (q, r) and $(\overline{q}, \overline{r})$ respectively. It has been shown by Li [20] that there exists two simple and elementary gauge matrices

$$T_1 = \begin{pmatrix} \lambda - \lambda_1 - \frac{1}{4}q\overline{r} & \frac{i}{2}q \\ \frac{i}{2}\overline{r} & 1 \end{pmatrix}, \quad T_2 = \begin{pmatrix} 1 & -\frac{i}{2}\overline{q} \\ -\frac{i}{2}r & \lambda - \lambda_2 - \frac{1}{4}\overline{q}r \end{pmatrix} \tag{4.2}$$

such that $\overline{\varphi} = T\varphi$, $T = T_1$ or T_2 and

$$\overline{q} - \frac{i}{2}q_x + \lambda_1 q + \frac{1}{4}q^2\overline{r} = 0$$
$$r + \frac{i}{2}\overline{r}_x + \lambda_1\overline{r} + \frac{1}{4}q\overline{r}^2 = 0 \tag{4.3}$$

and

$$q - \frac{i}{2}\overline{q}_x + \lambda_2 r + \frac{1}{4}\overline{q}^2 r = 0$$
$$\overline{r} + \frac{i}{2}r_x + \lambda_2 r - \frac{1}{4}\overline{q}r^2 = 0 \tag{4.4}$$

where λ_1 and λ_2 are arbitrary constants. Equations (4.3) and (4.4) are called the Backlund transformations (BTs) which transform one solution (q,r) to another $(\overline{q}, \overline{r})$ of the same equation in the AKNS hierarchy. The alternate application of (4.3) and (4.4) and their commuting property leads to the superposition formula

$$q_3 - q_0 = \frac{(\lambda_2 - \lambda_1)q_2}{1 - \frac{1}{4}q_2 r_1}, \quad r_3 - r_0 = \frac{(\lambda_1 - \lambda_2)r_1}{1 - \frac{1}{4}q_2 r_1} \tag{4.5}$$

namely, start from (q_0, r_0), we obtain (q_1, r_1) and (q_2, r_2) by (4.3) and (4.4) respectively, the next solution (q_3, r_3) can then be obtained algebraically by (4.5) without integrations.

Notice that the BT in (4.3) or (4.4) is still nonlinear. To avoid the nonlinearity, we observe that if φ and $\overline{\varphi}$ are matrix solutions, then $det\varphi$ and $det\overline{\varphi}$ are all constants. For a gauge matrix T being the polynomial in λ, $\overline{\varphi} = T\varphi$ implies that detT is also constant. As the polynomial, we assume that detT has simple zeros $\lambda_1, ..., \lambda_N$. At each point $\lambda = \lambda_j$, $det\overline{\varphi}(\lambda_j) = 0$, so constants exist such that the linear combination of rows of $\overline{\varphi}(\lambda_j) = T(\lambda_j)\varphi(\lambda_j)$ vanishes and therefore the coefficients of T can be solved in terms of $\varphi(\lambda_j)$, j=1,...,N and the combination constants.

In this way we find a simple gauge matrix T as follows

$$T = \begin{pmatrix} \lambda - \lambda_1 + \dfrac{(\lambda_2 - \lambda_1)\xi_1}{\xi_2 - \xi_1} & -\dfrac{\lambda_2 - \lambda_1}{\xi_2 - \xi_1} \\ \dfrac{(\lambda_2 - \lambda_1)\xi_1\xi_2}{\xi_2 - \xi_1} & \lambda - \lambda_2 - \dfrac{(\lambda_2 - \lambda_1)\xi_2}{\xi_2 - \xi_1} \end{pmatrix} \tag{4.6}$$

and the new solution is given by

$$\bar{q} = q - \frac{2i(\lambda_2 - \lambda_1)}{\xi_2 - \xi_1}, \quad \bar{r} = r - \frac{2i(\lambda_2 - \lambda_1)\xi_1\xi_2}{\xi_2 - \xi_1} \tag{4.7}$$

where

$$\xi_i = \frac{b_j\varphi_{21}(\lambda_j) + c_j\varphi_{22}(\lambda_j)}{b_j\varphi_{11}(\lambda_j) + c_j\varphi_{12}(\lambda_j)}, \quad j = 1, 2 \tag{4.8}$$

and $\varphi = (\varphi_{ij})_{2\times2}$ is regular, b_j, c_j, and λ_j are constants. So the new solution $(\bar{q}, \bar{r})$ is obtained as follows. Solve the linear system $\varphi_x = M\varphi$, $\varphi_t = N\varphi$ for the given solution (q, r) to get the regular matrix solution φ, then to obtain $(\bar{q}, \bar{r})$ by (4.7) algebraically. We call (4.7) the Darboux transformation (DT). By the DT, one only needs to solve a first order linear differential system and the remain steps for constructing the new solutions are purely algebraical. The discussion on the BT started at the beginning of eighties by one of the author (Y. S. Li) and a manuscript of [20] for BT of the AKNS system in detail was completed at that time, which was published in 1989 because of unknown reason. Details for DT can be found in [21] and the solution having the characteristics of homoclinic orbit for the NS equation obtained by the application of DT can be found in [22].

Further studies and application of DT for soliton equations, in particular for these in 2+1 dimensions were carried out by Gu, Hu and Zhou [23, 24, 25].

5 Symmetries for soliton equations

For a nonlinear evolution equation

$$u_t = K(u) \tag{5.1}$$

A symmetry $\sigma(u)$ is defined by the equation

$$\sigma_t(u) = K'[\sigma] \tag{5.2}$$

where

$$K'[\sigma] = \lim_{\varepsilon \to 0} \frac{\partial}{\partial \varepsilon} K(u + \varepsilon\sigma)$$

is the Frechet derivative. The symmetry corresponds to an infinitesimal symmetry transformation which leaves the solution space of (5.1) invariant when $u \to u + \varepsilon\sigma$. If the evolution equation

$$\frac{\partial\bar{u}}{\partial\varepsilon} = \sigma(\bar{u}), \quad \bar{u}|_{\varepsilon=0} = u \tag{5.3}$$

is solvable for u satisfying (5.1), we have then a one-parameter group of transformation [26]

Since 1985, we have investigated the topics of symmetries for soliton equations in both 1+1 and 2+1 dimensions and constructed two hierarchies of symmetries for many soliton equations. As for the generalized NS equation (2.11), the symmetries are [27]

$$a_n(u), \quad \tau_n(u) = 2ta_{n+1}(u) + b_n(u) \tag{5.4}$$

with a_n and b_n given in (2.7) and (2.8). These symmetries satisfy the following algebraic relations

$$\|[a_m, a_n]\| = 0, \quad \|[a_m, a_n]\| = ma_{n+m-1}$$
$$\|[\tau_m, \tau_n]\| = (m-n)\tau_{m+n-1} \tag{5.5}$$

where for any a(u) and b(u), the bracket is defined by

$$\|[a, b]\| = a'[b] - b'[a] \tag{5.6}$$

and $a'[b]$ etc. is the Frechet derivative. The derivation of these symmetries and the relations (5.5) is relied on the geometrical property of the recursion operator of the AKNS system.

It was noticed in [27] that with $\sigma(u)$ is either $a_n(u)$ or $\tau_n(u)$, the evolution equation (5.3) are nothing but the isospectral or non-isospectral flows associated with the AKNS spectral problem (2.6). Thus the initial value problem (5.3) for $\sigma = a_n$ or $\sigma = \tau_n$ is solvable via the IST.

For the 2+1 dimensional soliton equations. The construction of symmetries in [11, 12, 28] is based on the so-called Lax algebra rather than the extended recursion operator or the mastersymmetry. In the case of KP equation, the construction is as follows. First we derive two hierarchies of Lax operators A_n and B_n by the algebraically recursive formulae (2.20) and (2.21). For the potential being rapidly decreased as $x, y \to \infty$, the Lax operators satisfy

$$\|[A_m, A_n]\| = 0, \quad \|[A_m, B_n]\| = \frac{m+1}{3} A_{m+n-2},$$
$$\|[B_m, B_n]\| = \frac{m-n}{3} B_{m+n-2} \tag{5.7}$$

where the bracket is defined in (2.22). (5.7) implies that the linear space $\mathcal{A}$ spanned by A_n and B_n is a Lie algebra which is called the Lax algebra [11, 12]. As the consequence, we find that $K_n(u)$ and $\sigma_n(u)$ given by (2.18) and (2.19) also satisfy the same relations

$$\|[K_m, K_n]\| = 0, \quad \|[K_m, \sigma_n]\| = \frac{m+1}{3} K_{m+n-2},$$
$$\|[\sigma_m, \sigma_n]\| = \frac{m-n}{3} \sigma_{m+n-2} \tag{5.8}$$

but with the bracket of (5.6).

For the KP equation $u_t = K_2(u)$ in (2.14), the symmetries can be derived from (5.8), they are

$$K_n(u), \quad \tau_n(u) = tK_n(u) + \sigma_n(u) \tag{5.9}$$

with K_n and σ_n given in (2.18) and (2.19). The initial value problems (5.3) for $\sigma = K_n(u)$ and $\sigma = \tau_n(u)$ are solvable respectively by IST of the isospectral and non-isospectral deformations of the spectral problem (2.15).

We finally give the following remarks.

1) Based on the Lax algebra, one can systematically construct symmetries for many 2+1 dimensional soliton equations even for these possessing non-hereditary extended recursion operators (such as the system associated with the Schrodinger spectral problem in the plane). Detail can be found in [12, 28].

2) The Lax algebra also exists for the equations in 1+1 dimensions. For the AKNS hierarchy, the Lax algebra is spanned by A_n and B_n where A_n and B_n are in (2.7) and (2.8) (see [3]).

3) The Lax algebra implies the communitative properties for the infinitesimal symmetry transformations as well as for the one-parameter group of transformations generated by symmetries.

6 Summary

We have introduced several aspects on solitons and exactly solvable nonlinear evolution equations (soliton equations) carried out by us and our collaborators in the last decade. Because some of papers dealing with these aspects were published in the domestic journals and some of them even appeared in Chinese, we hope the present article is introductive to our readers. We apologize for omitting many of related literatures which should be quoted as references in the article.

7 References

[1] N. J. Zabusky and M. Kruskal, Phys. Rev. Lett. 15(1965), 240.

[2] C. Gardner, J. Greene, M. Kruskal and R. Miura, Phys. Rev. Lett. 19(1967), 1095.

[3] Y. Cheng and Y. S. Li, Chinese Science Bulletin 36 (1991), 1428.

[4] M. Ablowitz, D. Kaup, A. Newwell and H. Segur, Stud. Appl. Math. 53(1974), 249.

[5] H. H. Chen and S. C. Liu, Phy. Rev. Lett. 37(1976), 693.

[6] Y. S. Li, Scientia Sinica (Science in China) A25(1982), 911: Proc. of the 1980 Beijing Sym. on differetial geometry and differential equations, S. S. Chen W. T. Wu eds. (Science Press, Beijing 1980), V.3, 129.

[7] Y. S. Li, Ann. Of Math. (Chinese Edition) 2(1981), 147.

[8] D. W. Zhuang and Y. S. Li, Acta Math. Sinica 1(1985), 55.

[9] Y. Cheng, Y. S. Li and R. K. Bullough, J. Phy. A: Math. Gen. 21(1988), L433.

[10] Y. Cheng, Phy. Lett. A 127(1988),205.

[11] Y. Cheng, Physica D 46(1990), 286.

[12] Y. Cheng, in "Nonlinear Physics", Research Reports in Physics. Gu Chao-hao et. al. eds. (Springer-Verlag 1990), 12.

[13] M. Ablowitz and J. Ladik, J. Math. Phys. 16(1975), 598; ibid 17(1976), 1011.

[14] Y. Cheng, Scientia Sinica (Science in China) A 29(1986), 582.

[15] V. E. Zakharov and L. A. Takhtajan, Theor. Math. Phys. 38(1979), 26.

[16] Y. S. Li and D. W. Zhuang, Scientia Sinica (Science in China) A 26(1983), 811.

[17] D. Y. Chen, Y. S. Li and Y. B. Xeng, Scientia Sinica (Science in China), A 28(1985), 907.

[18] Y. S. Li, D. Y. Chen and Y. B. Zeng, in Proc. of DD4 S. T. Liao ed. (Science Press, Beijing 1986), 359.

[19] Y. Cheng, Y. S. Li and G. X. Tang, J. Phys. A: Math. Gen. 23(1990), L473.

[20] Y. S. Li, Adv. in Math. (Chinese Edition) 18(1989), 356(in Chinese).

[21] Y. S. Li, X. S. Gu and M. R. Zhou, Acta Math. Sinica (New Series) 3(1987), 143.

[22] Y. S. Li and C. Q. Wang, Kexue Tongbao (Chinese Science Bulletin) 16 (1985), 1209 (in Chinese).

[23] C. H. Gu, Lett. Math. Phys. 11(1986), 31.

[24] C. H. Gu and H. S. Hu, Lett. Math. Phys. 11(1986), 325.

[25] Z. X. Zhou, in "Nonlinear Phys.", Research Reports in Physics. Gu Chao-hao et. al. eds. (Springer-Verlag 1990), 23.

[26] C. Tian, in "Soliton Theory and its Application" Gu Chao-hao et. al. eds. (Zhejiang Science Press 1988). (English edition will be published by Springer-Verlag).

[27] Y. S. Li and G. C. Zhou, J. Phys. A: Math. Gen. 19(1986), 3713.

[28] Y. Cheng, J. Math. Phys. 32(1991), 157.

The Systems of Second Order Partial Differential Equations with Constant Coefficients [†]

Wei Lin and Ciquian Wu

Zhongshan University

1 Introduction—Definition of ellipticity of a system of PDEs

Consider the system [1]

$$A\frac{\partial^2 u}{\partial x^2} + 2B\frac{\partial^2 u}{\partial x \partial y} + C\frac{\partial^2 u}{\partial y^2} = f, \qquad (I)$$

where A, B, C are 2×2 real constant matrices and $u = (u_1, u_2)^T$, $f = (f_1, f_2)^T$ are two dimensional real vector functions of x, y.

System (I) is called to be elliptic in the sence of I.G. Petrovsky if

$$|A\alpha^2 + 2B\alpha\beta + C\beta^2| \neq 0,$$

for any real number α, β, $\alpha^2 + \beta^2 \neq 0$. In 1948 A.V. Bitsadze [2] gave two examples to show that the uniqueness of the solution to the Dirichlet problems for the elliptic system (in Petrovsky's sence) in any bounded closed domain cannot, in general, hold. Thus the Somigliana condition which guarantee that the Dirichlet problem of system (I) has a unique solution was introduced. In 1951, M.I. Visik [3] defined the concept of the strongly ellipticity for system (I) and proved the uniqueness of the solution to the Dirichlet problem for the strongly elliptic system. In [1] examples are provided to show that both the Somigliana condition and the Visik condition are sufficient but not necessary for the uniqueness of the corresponding Dirichlet problem. Therefore Professor Wu Xinmou presented the following problem: for the elliptic system in Petrovsky's definition, what is the necessary and sufficient condition for the uniqueness of the solutions to the Dirichlet problem in the arbitrary closed bounded domain? In 1960, Din Sjasi et al [1] proved:

[†] 1991 Mathematics Subject Classification: 35J55, 35L55, 35M05
[†] Supported by NNSF of China

C. Gu et al. (eds.), Partial Differential Equations in China, 173–181.
© 1994 Kluwer Academic Publishers.

Theorem: The Dirichlet problem for the elliptic system (I) has, at most, one solution in arbitrary closed bounded domain if and only if the coefficients of (I) satisfy the following condition:

$$|A + 2Bb + Cc| \neq 0, \quad \text{for any real } b, c, \text{ with } b^2 \leq c, \qquad (D)$$

This theorem concludes: for the system (I), the necessary condition for the uniqueness of Dirichlet problem for the domain bounded by arbitrary closed second order curve is also the sufficient condition for the arbitrary closed bounded domain. So, it seems to be more suitable to consider condition (D) as the definition of ellipticity of system (I). This theorem is mainly proved by constructing a nonsingular matrix K such that system (I) becomes strongly elliptic when multiplied on the left by K.

Din's work is a very excellent result. In 1962, Professor Hua Loo Keng et al revised the definition of the Visik's condition and of the Somigliana's condition and proved that they are equivalent to condition (D). Further he suggested Wu Ci-Quian and Lin Wei to classify the linear system into different types and systematically research each type separately. Below the main result on this topic will be presented.

2   Canonical form and classification

Definition 2.1 Sytem

$$\left[A\frac{\partial^2}{\partial x^2} + 2B\frac{\partial^2}{\partial x \partial y} + C\frac{\partial^2}{\partial y^2} \right] \begin{pmatrix} u \\ v \end{pmatrix} = 0, \qquad (I)$$

and system

$$\left[A_1\frac{\partial^2}{\partial x^2} + 2B_1\frac{\partial^2}{\partial x \partial y} + C_1\frac{\partial^2}{\partial y^2} \right] \begin{pmatrix} u \\ v \end{pmatrix} = 0, \qquad (I_1)$$

are said to be equivalent if one can be transformed into the other by means of successive applications of the following three kinds of operation: (i) linear combination of equations, (ii) linear transformation of unknown functions, (iii) linear transformation of independent variables.

Definition 2.2 The determinant $Q(\xi, \eta) = |A\xi^2 + 2B\xi\eta + C\eta^2|$ is called the biquadratic characteristic form of sytsem (I).

It is easy to prove:

Theorem 2.1   Operation (i) and (ii) carry the biquadratic characteristic form $Q(\xi, \eta)$ of system (I) into itself apart from a nonzero constant multiple $|P| \cdot |Q|$. If operation (iii) $\begin{pmatrix} x_1 \\ y_1 \end{pmatrix} = \begin{pmatrix} p & q \\ r & s \end{pmatrix} \begin{pmatrix} x \\ y \end{pmatrix}$ is applied to system (I), then $Q_1(\xi, \eta) = |A_1\xi^2 + 2B_1\xi\eta + C_1\eta^2| = F(\xi', \eta')$, where $\xi' = p\xi + r\eta$, $\eta' = q\xi + s\eta$.

Let $\xi_i/\eta_i = \tau_i$, $\xi_i'/\eta_i' = \tau_i'$ (i=1,2,3,4) denote roots of $Q_1(\tau, 1) = 0$ and $Q(\tau', 1) = 0$ respectively. Then

$$\tau' = \frac{p\tau_i + r}{q\tau_i + s}, \qquad i = 1, 2, 3, 4 \tag{2.1}$$

This means that a linear transformation with real coefficients of the characteristic roots τ will correspond to the operation (iii).

The biquadratic form has nine different combinations of roots. By using a real linear transformation (2.1), the biquadratic characteristic form may be reduced into the standard form. For example, when $Q(\tau, 1) = 0$ has two distinct pairs of complex roots. Let τ_1, $\bar{\tau}_1$, τ_2, $\bar{\tau}_2$ be such four complex roots. We may draw a circle passing through points τ_1, τ_2, $\bar{\tau}_1$, $\bar{\tau}_2$ and also orthogonally cutting the real linear transformation $w = \rho\frac{z-p}{z-q}$ maps this circle onto the imaginary axis and maps points τ_1, $\bar{\tau}_1$, τ_2, $\bar{\tau}_2$ into $ki, -ki, i, i, 0 < k < 1$, where ρ is a real number. This implies $Q(\xi, \eta) = (\xi^2 + \eta^2)(\xi^2 + k^2\eta^2)$, $0 < k < 1$. So we have

Theorem 2.2 According to the properties of its roots, the biquadratic characteristic form may be reduced by means of operation (iii) into one of the following two standard forms:

$$Q(\xi, \eta) = (\xi^2 + \varepsilon\eta^2)(\xi^2 + K^2\eta^2), \tag{III}$$

where

$$\varepsilon = 1, \begin{cases} 0 < k < 1, \text{ for } Q(\xi, \eta) \text{ having two distinct pairs of complex roots;} \\ k=1, \text{ for } Q(\xi, \eta) \text{ having a pair of double complex roots;} \\ k=0, \text{ for } Q(\xi, \eta) \text{ having a pair of complex roots and a} \\ \text{double real roots;} \end{cases}$$

$\varepsilon = 0$, k=0, $Q(\xi, \eta)$ having a quadruple rel root;

and

$$Q(\xi, \eta) = \xi\eta(\delta\xi^2 + 2\alpha\xi\eta + \varepsilon\eta^2), \tag{IV}$$

where

$$\delta = \varepsilon = 1, \begin{cases} 0 \leq \alpha < 1, \text{ for } Q(\xi, \eta) \text{ having a pair of complex} \\ \text{and two distinct real roots;} \\ \alpha = 1, \text{ for } Q(\xi, \eta) \text{ having three distinct real roots;} \\ \alpha > 1, \text{ for } Q(\xi, \eta) \text{ having four distinct real roots;} \end{cases}$$

$\delta = 1$, $\varepsilon = \alpha = 0$, for $Q(\xi, \eta)$ having a triple and a simple real root;
$\delta = \varepsilon = 0$, $\alpha = 1$, for $Q(\xi, \eta)$ having two double real roots.

Starting from the standard form (III) and (IV), applying the operations (i) and (ii), we may reduce the system (I) into the canonical form. It is easy to prove:

Theorem 2.3 The system (I), possessing characteristic biquadratic form (III), is equivalent to the following canonical form:

$$A = \begin{pmatrix} 1 & 0 \\ 0 & 1 \end{pmatrix}, \quad B = \begin{pmatrix} 0 & 1 \\ b & 0 \end{pmatrix}, \quad C = \begin{pmatrix} \lambda & 0 \\ 0 & \mu \end{pmatrix}, \tag{A}$$

$$\lambda + \mu - 4b = k^2 + \varepsilon, \quad \lambda\mu = k^2\varepsilon, \quad b \neq 0, \quad (b < 0, \quad if \ \lambda = \mu)$$

Theorem 2.4 The system (I), possessing characteristic biquadratic form (IV), is equavilent to following:

$$A = \begin{pmatrix} 1 & 0 \\ 0 & 0 \end{pmatrix}, \quad B = \begin{pmatrix} \frac{\varepsilon}{2c} & 1 \\ b & \frac{\delta}{2c} \end{pmatrix}, \quad C = \begin{pmatrix} 0 & 0 \\ 0 & 1 \end{pmatrix}, \qquad (B)$$

$$b = \frac{\delta\varepsilon}{4c^4} - \frac{\alpha}{2c} + \frac{1}{4} \neq 0, \quad c \neq 0$$

System (A) and (B) may be classified according to the definitions given by Petrovsky for the classification of systems of PDEs [6].

Definition 2.3 System (A) $Q(\xi, \eta)$, possessing a pair of double complex roots is called the elliptic system of the first kind and denoted by (E_1). System (A) with $Q(\xi, \eta)$, possessing two pairs of complex roots is called the elliptic system of the second kind and denoted by (E_2).

Definition 2.4 System (A) with $Q(\xi, \eta)$ having a pair of complex roots and a double real root is called the composite system of the first kind and denoted by (C_1). System (B) with $Q(\xi, \eta)$ having a pair of complex and two distinct real roots is called the composite system of the second kind and denoted by (C_2).

Definition 2.5 System (B) is referred to as the hyperbolic system of the first, second, third and fourth kinds in the cases $\delta = \varepsilon = 1, \ \alpha > 1; \ \delta = \varepsilon = 1, \ \alpha = 1; \ \delta = \varepsilon = 0, \ \alpha = 1; \ \delta = 1, \ \varepsilon = \alpha = 0$ respectively. They are denoted by (H_1), (H_2), (H_3), (H_4) correspondingly.

Definition 2.6 System (A) with $Q(\xi, \eta)$ having a quadruple real root is called parabolic and denoted by (P).

In [9] the classification of system (I), including the case $Q(\xi, \eta) \equiv 0$, in associate with its lower terms is discussed.

Starting from canonical form, it is more simple to prove the uniqueness theorem of Dirichlet problem for elliptic systems [7]. Further we may systematically [6, 8] investigate each type.

3 Bi-analytic functions of type (λ, k) and their applications in elasticity

J. Sander [10] discussed the function class determined by the system of PDEs of first order and introduced the concept of bi-analytic functions. The bi-analytic function connected with system (E_2) is considered in [11]. We may rewrite system (E_2) in the form:

$$\left[\begin{pmatrix} k & 0 \\ 0 & -\lambda \end{pmatrix}\frac{\partial}{\partial x} + \begin{pmatrix} 0 & \lambda \\ k & 0 \end{pmatrix}\frac{\partial}{\partial y}\right]\left[\begin{pmatrix} \frac{1}{k} & 0 \\ 0 & \frac{1}{k} \end{pmatrix}\frac{\partial}{\partial x} + \begin{pmatrix} 0 & -1 \\ 1 & 0 \end{pmatrix}\frac{\partial}{\partial y}\right]\begin{pmatrix} u \\ v \end{pmatrix} = 0$$

$$0 < k < 1, \quad \lambda \neq 0, 1, k^2 \qquad (E_2)$$

If u, v are differentiable up to the second order, the elliptic system (E_2) is equivalent to the first order system:

$$\begin{cases} \frac{1}{k}\frac{\partial u}{\partial x} - \frac{\partial v}{\partial y} = \theta, \\ \frac{\partial u}{\partial y} + \frac{1}{k}\frac{\partial v}{\partial x} = \omega, \end{cases} \qquad \begin{cases} k\frac{\partial \theta}{\partial x} + \lambda\frac{\partial \omega}{\partial y} = 0, \\ k\frac{\partial \theta}{\partial y} - \lambda\frac{\partial \omega}{\partial x} = 0, \end{cases} \qquad (K_\lambda)$$

$$\lambda \neq 0, 1, k^2, \quad 0 < k < 1.$$

Definition 3.1 The function f(Z)=u+iv is called the bi-analytic function of type (λ, k) if u, v are the solutions of the system

$$\left\{ \frac{1}{k}\frac{\partial u}{\partial x} - \frac{\partial v}{\partial y} = \theta, \frac{\partial u}{\partial y} + \frac{1}{k}\frac{\partial v}{\partial x} = \omega, \right. \qquad (3.1)$$

where the functions θ, ω in turn satisfy the equations

$$\left\{ k\frac{\partial \theta}{\partial x} + \lambda\frac{\partial \omega}{\partial y} = 0, k\frac{\partial \theta}{\partial y} - \lambda\frac{\partial \omega}{\partial x} = 0, \right. \qquad (3.2)$$

with λ, k real constants and $\lambda \neq 0$, 1, k^2, $0 < k < 1$. Evidently, function $\phi(z) = k\theta - i\lambda\omega$ is analytic, and we refer to $\phi(z)$ as the associate function of f(z).

In [11], definitions of integral, derivative, generalized power functions for the bi-analytic function are introduced and the function theory, e.g. Cauchy integral theorem and formula, Taylor and Laurent series etc., is also developed. Moreover, Cauchy type integral and Plemej's formula are rewritten as the complex form

$$\frac{k+1}{2k}\frac{\partial f}{\partial \overline{Z}} - \frac{k-1}{2k}\frac{\partial f}{\partial Z} = \frac{\lambda-k}{4\lambda}\phi(Z) + \frac{\lambda+k}{4\lambda}\overline{\phi(Z)},$$

and consequently we obtain the representation of the general solution to system (E_2) [8]

$$f(Z) = \frac{\lambda-k}{2(1-k)\lambda}\Phi(Z) + \frac{\lambda+k}{2(1+k)\lambda}\overline{\Phi(Z)} + \Psi(\frac{k+1}{2k}Z + \frac{k-1}{2k}\overline{Z}), \qquad (3.3)$$

where $\Phi(Z) = \int_{Z_0}^{Z} \phi(\varsigma)d\varsigma$.

Some boundary value problems for system (E_1) and (E_2) are solved by using the theory of bi-analytic function [8, 12, 13]. As is known, the theory of analytic functions has many applications in elasticity, it is thus natural to consider the problem of plane elasticity by using the above results. For plane strain, the Navier equations for an orthotripic body, without forces become

$$
\left[\left[\begin{pmatrix} E_{11} & 0 \\ 0 & G_{12}(1-\nu_{12}\nu_{21}) \end{pmatrix} \frac{\partial^2}{\partial x^2} + (\nu_{12}E_{22} + G_{12}(1-\nu_{12}\nu_{21}) \right. \right.
$$

$$
\left. \left. \begin{pmatrix} 0 & 1 \\ 1 & 0 \end{pmatrix} \frac{\partial^2}{\partial x \partial y} + \begin{pmatrix} G_{12}(1-\nu_{12}\nu_{21}) & 0 \\ 0 & E_{22} \end{pmatrix} \frac{\partial^2}{\partial y^2} \right] \begin{pmatrix} u \\ v \end{pmatrix} = 0, \right. \tag{3.4}
$$

where ν_{12}, ν_{21} — the Poisson ratios, E_{11}, E_{22} — the Young's moduli, and $\nu_{12}E_{22} = \nu_{21}E_{11}$. We may prove [14]

Lemma: The Navier equations of plane strain for an orthotropic body may, by means of the composite transformations of coordinates and unknown function $x \to \delta x, y \to y/\sqrt{k_1}$ and $u \to \frac{1+k_1\nu}{k_1+\nu}\frac{1}{\sqrt{k_1}}u, v \to \delta v$, be put into canonical form (E_2), where $\nu^2 = \nu_{12}\nu_{21}, E^2 = E_{11}E_{22}, \delta^2 = \sqrt{\frac{E_{11}}{E_{22}}}, k_1 = \frac{E}{2G_{12}} - \nu, k = k_1 - \sqrt{k_1^2 - 1}(k > 1),$ $\lambda = \frac{1-\nu^2}{2(k_1+\nu)}k.$

This lemma shows that the system of the Navier equations for the orthotropic body is equivalent to system (E_2), so we can utilize the results about the bi-analytic function of type (λ, k) to solve the basic problem —— the displacement boundary value problem and the stress boundary value problem [14].

4   The hyperbolic system

The canonical forms of four classes of the hyperbolic systems are [4 - 8, 15, 16].

$$
\left[\begin{pmatrix} 1 & 0 \\ 0 & 0 \end{pmatrix} \frac{\partial^2}{\partial x^2} + 2 \begin{pmatrix} b_1 & 1 \\ b_3 & b_1 \end{pmatrix} \frac{\partial^2}{\partial x \partial y} + \begin{pmatrix} 0 & 0 \\ 0 & 1 \end{pmatrix} \frac{\partial^2}{\partial y^2} \right] \begin{pmatrix} u \\ v \end{pmatrix} = 0, \tag{H_1}
$$

$$
b_3 = b_1^2 - \frac{1}{2}(k + \frac{1}{k})b_1 + \frac{1}{2} \neq 0, \quad b_1 \neq 0, \quad 0 < k < 1;
$$

$$
\left[\begin{pmatrix} 1 & 0 \\ 0 & 0 \end{pmatrix} \frac{\partial^2}{\partial x^2} + 2 \begin{pmatrix} b_1, & 1 \\ (b_1+\frac{1}{2})^2 & b_1 \end{pmatrix} \frac{\partial^2}{\partial x \partial y} + \begin{pmatrix} 0 & 0 \\ 0 & 1 \end{pmatrix} \frac{\partial^2}{\partial y^2} \right] \begin{pmatrix} u \\ v \end{pmatrix} = 0,
$$
$$
\tag{H_2}
$$

$$
b_1 \neq \frac{1}{2}
$$

$$
\left[\begin{pmatrix} 1 & 0 \\ 0 & 0 \end{pmatrix} \frac{\partial^2}{\partial x^2} + 2 \begin{pmatrix} 0 & 1 \\ \frac{1}{4} - k & 0 \end{pmatrix} \frac{\partial^2}{\partial x \partial y} + \begin{pmatrix} 0 & 0 \\ 0 & 1 \end{pmatrix} \frac{\partial^2}{\partial y^2} \right] \begin{pmatrix} u \\ v \end{pmatrix} = 0, \tag{H_3}
$$

$$
k \neq 0, \frac{1}{4};
$$

$$\left[\begin{pmatrix} 1 & 0 \\ 0 & 0 \end{pmatrix}\frac{\partial^2}{\partial x^2} + 2\begin{pmatrix} 0 & 1 \\ \frac{1}{4} & k \end{pmatrix}\frac{\partial^2}{\partial x \partial y} + \begin{pmatrix} 0 & 0 \\ 0 & 1 \end{pmatrix}\frac{\partial^2}{\partial y^2}\right]\begin{pmatrix} u \\ v \end{pmatrix} = 0, \qquad (H_4)$$

$$k \neq 0$$

Assume $y = cx$ not to be the characteristic line of systems (H_i) $(i=1,2,3,4)$. We may prove that there exists a unique solution to the Cauchy problem for (H_i), the data of which u, v and $\frac{\partial u}{\partial n}$, $\frac{\partial v}{\partial n}$ are given on the support $y=cx$.

Suppose l_j to be the characteristic line of systems (H_i) $(i=1,2,3,4)$. The so-called characteristic problem is to seek out the solution to the systems (H_i) such that on l_j they take the values of the given functions. We have the following results: i) The characteristic problem for system (H_4) does not, in general, exist; ii) the solution to the characteristic problem for system (H_3) is unique; iii) the solution to the characteristic problem for system (H_1) and system (H_2) is, in general, not unique, however, the necessary and sufficient condition of its non-uniqueness is found out and the representation of the non-unique solution is obtained.

The investigation of above problems is reduced to the following theorem [17-19].

Theorem: The homogeneous functional equation

$$f(x) = \sum_{i=1}^{l} a_i f(\alpha_i, x), \quad |\alpha_i| < 1, \quad -\infty < x < +\infty, \qquad (F)$$

only has the solution, being identically equal to a constant, in the function class $\mathcal{F}_1 = \{f(x)|f(x) \in C^1(U(0)), \quad f'(x) = O(|x|^{\mu-1})$ as $x \to 0$, where $\mu \geq 1$ such that the real functions X^μ and $X^{\mu-1}$ are well defined for all real x and $\sum_{i=1}^{l}|a_i||\alpha_i|^\mu < 1\}$. In the function class $\mathcal{F}_2 = \{f(x)|f(x) \in C^1(U(0)), \quad \lim_{x \to 0} x^{1-\mu}f'(x)$ exists, where $\mu \geq 1$, $\sum_{i=1}^{l}|a_i||\alpha_i|^\mu = 1$ and such that the real functions x^μ and $x^{\mu-1}$ are well defined $\}$, equation (F) has the solution $ax^{mu} + b$, where

$$a = \begin{cases} 0, & if \ \sum_{i=1}^{l} a_i\alpha_i^\mu \neq 1, \\[2mm] arbitrary \ \ real \ \ number, & if \ \sum_{i=1}^{l} a_i\alpha_i^\mu - 1. \end{cases}$$

$$b = \begin{cases} 0, & if \ \sum_{i=1}^{l} a_i \neq 1, \\[2mm] arbitrary \ \ real \ \ number, & if \ \sum_{i=1}^{l} a_i = 1. \end{cases}$$

Xuan Qiwo generalized above theorem to the case of the functional matrix equation [20].

Wu Ci-Quian etc. discussed the iteration method, location solution, Galerkin method, Monteo Carlo method for solving the functional equations and applied it to obtain the numerical solution of the hyperbolic systems [2]-[24].

Xuan Qiwo also investigated the influence of the lower order terms on solutions to systems (H_1), (H_2), (H_3) [20].

5   The complete systems

The complete system may be reduced to the canonical form

$$\left[\begin{pmatrix} 1 & 0 \\ 0 & -1 \end{pmatrix} \frac{\partial}{\partial x} + \begin{pmatrix} 0 & 1 \\ 1 & 0 \end{pmatrix} \frac{\partial}{\partial y}\right] \left[\begin{pmatrix} 1 & 0 \\ 0 & 2 \end{pmatrix} \frac{\partial}{\partial x} + \begin{pmatrix} 0 & \frac{2k^2}{\lambda} \\ \lambda & 0 \end{pmatrix} \frac{\partial}{\partial y}\right] \begin{pmatrix} u \\ v \end{pmatrix} = 0,$$

$$\lambda \neq 0, \quad k \geq 0, \tag{C}$$

where $k = 0$ and $k > 0$ correspond to the first kind (C_1) and the second kind (C_2) respectively. Systems of composite type are unclassical, they possess some of the characteristic not only of elliptic systems but also of hyperbolic systems. So, there are more formulations of the boundary value problems, initial problems and various mixed problems to be considered [25-28].

References

[1] Ding Shia-kuai, Wang Kan-ting, Ma Ju-nien, Shun Chia-lo, Chang Tong, Definition of ellipticity of a system of second order partial differential equations with constant coefficients, Acta Math. Sinica, 10(1960), 276-287.

[2] A.V. Bitsadze, Some linear problems of linear partial differential equations, Advances in Math. (China), 3(1958), 321-399.

[3] M.I. Visik, Strongly elliptic systems of differential equations, Math. Sbornik, 29(1951), 613-676.

[4] Hua Loo-Keng, Wu Ci-Quian and Lin Wei, The canonical form of the system of partial differential equations of second order, Kexue Tongbao (Beijing), 16(1964), 1100-1103.

[5] Hua Loo-Keng, Wu Ci-Quian and Lin Wei, On classification of the system of differential equations of the second order, Scientia Sinica, 3(1965), 461-465.

[6] Hua Loo-Keng, Wu Ci-Quian and Lin Wei, Systems of second order linear partial differential equations with cnstant coefficients, two independent variables and two functions, Science Press, Beijing, China, 1979.

[7] Hua Loo-Keng, Wu Ci-Quian and Lin Wei, On the uniqueness of the solution of the Dirichlet problem of the elliptic system of differential equations, Acta Math. Sinica, 15(2), (1965).

[8] Hua Loo-Keng, Lin Wei and Wu Ci-Quian, Second order systems of partial differential equations in the plane, Research Notes in Math. No. 128, Pitman Publishing Inc., 1985.

[9] Wu Ciquian, Liu Yunkang, Xuan Qiwo, Lin Xingqing, On classification for the second order systems of partial differential equations with constant coefficients, Acta Scientiarum Naturalium Universitatis Sunyatseni, Supplement 1(1989).

[10] J. Sander, Viscous fluids elasticity and function theory, Trans. Amer. Math. Soc., 98(1961), 85-147.

[11] Lin Wei and Wu Ciquian, On the bi-analytic functions of (λ, k) types, Acta Scientiarum Naturalium Universitatis Sunyantseni, 1(1965).

[12] Lin Wei and Wu Ciquian, On the systems of the elliptic equations of second order, Acta Scientiarum Naturalium Universitatis Sunyatseni, 4(1964).

[13] Xu Yongzhi, Cauchy type integral of (λ, k) bi-analytic functions and Riemann problem, Acta Scientiarum Naturalium Universitatis Sunyatseni (Mathematics), 1(1984).

[14] R. P. Gilbert and Lin Wei, Function theoretic solutions to problems of orthotropic elasticity, J. Elasticity, 2(1985).

[15] Wu Ci-Quian and Lin Wei, The linear systems of hyperbolic equations of second order with constant coefficients (I), Acta Scientiarum Naturalium Universitatis Sunyatseni , 3(1964).

[16] Wu Ci-Quian and Lin wei, The linear systems of hyperbolic equations of second order with constant coefficints (II), Acta Scientiarum Naturalium Universitatis Sunyatseni, 4(1965).

[17] Wu Ci-Quian and Lin Wei, The first characteristic problem for systems of hyperbolic equation, Acta Scientiarum Naturalium Universitatis Sunyatseni, 2(1980).

[18] Lin Wei and Wu Ci-Quian, The second characteristic problem for systems of the hyperbolic equations, Acta Scientiarum Naturalium Universitatis Sunyatseni, 3(1980).

[19] Wu Ci-Quian and Lin Wei, The third characteristic problem for systems of hyperbolic equations, Acta Scientirum Naturalium Universitatis Sunyatseni, 4(1980).

[20] Wu Ci-Quian and Xuan Qiwo, Acta Scientiarum Naturalium Universitatis Sunyatseni, Supplement, 1(1989).

[21] Wu Ci-Quian, Acta Scientiarum Naturalium Universitatis Sunyatseni, 4(1980).

[22] Wu Ci-Quian and Xuan Qiwo, Acta Scientiarum Natiralium Universitatis Sunyatseni, 1(1984).

[23] Wu Ci-Quian and Xuan Qiwo, Acta Scientiarum Naturalium Universitatis Sunyatseni, 4(1986).

[24] Wu Ci-Quian and Xuan Qiwo, Acta Scientiarum Naturalium Universitatis Sunyatseni, Supplement, 1(1989), 68-71.

[25] Wu Ci-Quian and Lin Wei, On the systems of PDEs of the composite type of second order, Acta Scientiarum Naturalium Universitatis Sunyatseni, 4(1984).

[26] Lin Wei and Wu Ci-Quian, Some problems for the systems of equations composite type, Acta Scientiarum Naturalium Universitatis Sunyatseni, 3(1984).

[27] Lin Wei, Some second order systems of PDEs of Composite type, Proceedings of the Toyal Society of Edinburgh, 106A, 1(1987).

[28] Dai Daoqing, A boundary value problem for the composite system of second order equations, Acta Scientiarum Natiuralium Universitatis Sunyatseni, 2(1988).

Other *Mathematics and Its Applications* titles of interest:

D.S. Mitrinovic, J.E. Pecaric and A.M. Fink: *Classical and New Inequalities in Analysis.* 1992, 740 pp.   ISBN 0-7923-2064-6

H.M. Hapaev: *Averaging in Stability Theory.* 1992, 280 pp.   ISBN 0-7923-1581-2

S. Gindinkin and L.R. Volevich: *The Method of Newton's Polyhedron in the Theory of PDE's.* 1992, 276 pp.   ISBN 0-7923-2037-9

Yu.A. Mitropolsky, A.M. Samoilenko and D.I. Martinyuk: *Systems of Evolution Equations with Periodic and Quasiperiodic Coefficients.* 1992, 280 pp.
ISBN 0-7923-2054-9

I.T. Kiguradze and T.A. Chanturia: *Asymptotic Properties of Solutions of Non-autonomous Ordinary Differential Equations.* 1992, 332 pp.   ISBN 0-7923-2059-X

V.L. Kocic and G. Ladas: *Global Behavior of Nonlinear Difference Equations of Higher Order with Applications.* 1993, 228 pp.   ISBN 0-7923-2286-X

S. Levendorskii: *Degenerate Elliptic Equations.* 1993, 445 pp.
ISBN 0-7923-2305-X

D. Mitrinovic and J.D. Kečkić: *The Cauchy Method of Residues, Volume 2.* Theory and Applications. 1993, 202 pp.   ISBN 0-7923-2311-8

R.P. Agarwal and P.J.Y Wong: *Error Inequalities in Polynomial Interpolation and Their Applications.* 1993, 376 pp.   ISBN 0-7923-2337-8

A.G. Butkovskiy and L.M. Pustyl'nikov (eds.): *Characteristics of Distributed-Parameter Systems.* 1993, 386 pp.   ISBN 0-7923-2499-4

B. Sternin and V. Shatalov: *Differential Equations on Complex Manifolds.* 1994, 504 pp.   ISBN 0-7923-2710-1

S.B. Yakubovich and Y.F. Luchko: *The Hypergeometric Approach to Integral Transforms and Convolutions.* 1994, 324 pp.   ISBN 0-7923-2856-6

C. Gu, X. Ding and C.-C. Yang: *Partial Differential Equations in China.* 1994, 181 pp.   ISBN 0-7923-2857-4

The manufacturer's authorised representative in the EU is Springer
Nature Customer Service Centre GmbH, Europaplatz 3, 69115 Heidelberg,
Germany. If you have any concerns regarding our products, please
contact ProductSafety@springernature.com

Printed and bound by CPI Group (UK) Ltd, Croydon, CR0 4YY

23/04/2026

02095623-0003